KLONDIKE WOMEN

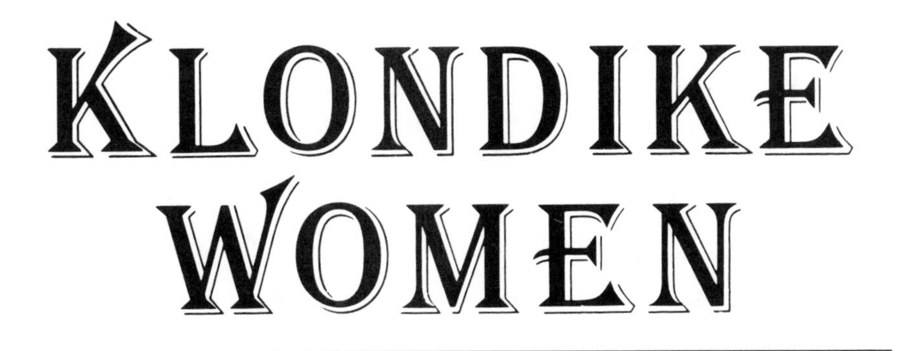

KLONDIKE WOMEN

True Tales of the 1897–98 Gold Rush

MELANIE J. MAYER

SWALLOW PRESS · OHIO UNIVERSITY PRESS

01 00 9 8 (pbk.)
97 96 95 94 93 92 7 6 5 4 3 2 (cloth)

Swallow Press / Ohio University Press books are printed on acid-free paper ∞

Library of Congress Cataloging-in-Publication Data

Mayer, Melanie, J., 1945–
Klondike Women : true tales of the 1897-98 Gold Rush /
Melanie J. Mayer.
p. cm. Bibliography : p. Includes index.
ISBN 0-8040-0926-0 (cloth). ISBN 0-8040-0927-9 (pbk.)
1. Klondike River Valley (Yukon)–Gold discoveries. 2. Women pioneers–
Yukon Territory–Klondike River Valley–Biography.
3. Frontier and pioneer life–Yukon Territory–Klondike River Valley. I. Title
F1095.K5M29 1989 971.9'1–dc20 89-33517 CIP

Designed by Laury A. Egan

. . . no one has ever loved an adventurous woman
as they have loved adventurous men.

ANAÏS NIN, NOVEMBER 1934

DEDICATION

For love of adventurous women,
that we may love our adventurous selves;
for Edith Feero Larson, 1887–1981.

CONTENTS

Acknowledgments ix

Introduction
"Gold! Gold! Gold! Gold!" 3

1 Gold Rush Beginnings
"Bring me a stocking full of gold" 13

2 Edmonton, The Poor Folk's Way
"The boys threw dishes at her" 28

3 Steaming In via the All-Water Yukon Route
"I came 13,000 miles without a cent, so I'm satisfied" 52

4 Trekking the Chilkoot Trail
"For God's sake, Polly, buck up and be a man" 64

5 Skagway, Gold Rush Boomtown
"I always wanted to try making my own way" 111

6 Packing on the White Pass Trail
"I can be a lady on the trail" 130

7 Shooting the Yukon Headwaters' Rapids
"I don't know when I enjoyed anything so much in my life" 167

8 Dawson, and then . . .
"What I wanted was liberty and opportunity" 196

Notes 237

References 252

Index 259

ACKNOWLEDGMENTS

OCCASIONALLY life affords one the opportunity to go exploring. And then, quite unexpectedly, the external, physical encounter with the unknown becomes a metaphor for internal, mental probing into new territories as well. For me one of those times came in 1979 when I hopped aboard the Alaska State Ferry in Seattle and wandered north. With nothing more than what I could carry on by back, I was off to see Alaska and the Yukon.

Alaska! I had heard my father's wartime stories of Fairbanks since I was a child. What about this land that is so close to Asia, a part of the United States and yet so far away, a land which had held my beginnings, at least for a time? *Yukon!* That deed for one square inch of the Klondike which I fished out of a cereal box must have touched something deep in my imagination. I still had it tucked away with other "important papers" from my childhood. Do I still own part of the Klondike gold? Now I had the opportunity to really see these places with which I had had such tenuous but intriguing connections for so many years. Perhaps I should have guessed that this journey might turn into something more than a casual vacation.

The impact of Alaska and the Yukon on me was and still is profound. The vast, natural beauty and power of the Northwest presses round me, sings to me, and demands respect. The land touches me physically in a way that I have very little hope of explaining. So I just note it—and I am awed. Yes, there is "civilized" development and, as has been true since the first Caucasian contacts, exploitation of the natural resources of the country. But what I love is the wilderness, the natural wonders of the place, and the people who lovingly make this country their home.

And on that first trip to the North, I was fascinated by photographs here and there showing women making their way into the country in the late 1800s. Who were they? What were Victorian women doing rushing to the Klondike for gold? They didn't look like the typical prostitutes and dance hall "girls" from Klondike stories I had read. And who were those Native Americans? Why hadn't I heard about all these women? And so the beginnings of another exciting exploration and discovery began to take shape—a book about the Klondike women. It has held my attention through the years, always delighting me as new information comes to light. It has introduced me to many new and interesting people. And it has tested and strengthened relationships with old friends.

I would like to thank publicly all those who have helped to make this book possible. And I extend special recognition to the following for their generous assistance:

EDITH LARSON's children, after Edith died, supplied missing information about her life and donated her photography collection to the University of Washington, Seattle, for scholarly use. So I thank Stewart and Helene Larson, Ranie DeHaven, and Ellen Matchett.

Victoria Faulkner, Whitehorse (formerly of Dawson); Evelyn Lee, Victoria (formerly of Mayo); Art and Margaret Fry, Dawson; Betty Mulrooney, Spokane; and Gladys Gue Wilfong and Marjorie Gue Lambert, Puyallup, gave informative interviews.

The staff at the University of Washington Libraries Special Collections Division, including the Northwest, Manuscripts, and Photography Collections, have done a first-rate job in making their material accessible and understandable. I also appreciate very much their personal kindness to me. Dennis Andersen and Richard Engeman, Curators of the Photography Collection, and Sandra Kroupa, Curator of Rare Books, were especially helpful.

Ruth Herberg, Seattle, allowed me to use her translation of Inga Sjolseth Kolloen's diary. The Kolloen family deposited Inga and Henry Kolloen's material with the University of Washington for use by scholars. Erling Kolloen was especially helpful with details about Inga's life.

My friends in Seattle and Fairbanks gave me lots of encouragement, good advice, and their homes to share while I did research for this book. Among these are Jill Schultz, Lee Cooper, Wayne and Chris Gustafson, Richard Frundt, Cindy Lyons, and Davida Teller. Christine Moss Dunkel has been especially generous, involved, and tolerant over the years. Thanks, Tina.

Peter Heebink went out of his way to welcome my research expedition to Whitehorse in the summer of 1980. With his help, I was able to experience firsthand the Yukon River between Whitehorse and Dawson. Also in Whitehorse that summer, Kathy Williams shared her house with warm, Yukon hospitality, and Stu Withers lent us his well-crafted canoe.

My Yukon and Alaskan traveling friends helped to create modern adventures in the North—Lee Cooper, Angie Kuper Christmann, Tina Moss Dunkel, Larry Dunkel, Jim Christmann, Barry Ziker, and Scott Ferguson.

The staff at the Yukon Archives, Whitehorse, including Marion Ridge and Diane Johnston, have been most generous with their assistance, as have India Spartz, Marilyn Kwock, and Kathryn Shelton at the Alaska

Historical Library, Juneau. Also helpful were Diane Brenner, Anchorage Historical and Fine Arts Museum; Keith Boettcher, Fresno County Free Library; Isabel Miller at the Centennial Museum, Sitka; and the library staff of the University of Alaska, Fairbanks.

Back in Santa Cruz, the Interlibrary Loan staff at the University of California helped me to do the necessary research-at-a-distance. I thank especially Betty Rentz for her efforts in locating hard-to-find documents.

Dorothy DeBoer, Kermit Edmonds, C. J. Peter Bennett, Jean King, Dale and Sarah Kroll, and Rose Vitzhum supplied some previously unpublished material about women on the Klondike trails. Marie Morris and Ella S. Michael were especially tenacious in helping to locate material that solved some puzzles.

Robert and Dale DeArmond welcomed me warmly to Juneau and, through their appreciation of Alaskan life and history, increased my own. In addition, Bob has helped dig out and generously shared many facts about the Klondike women.

Mort Yanow, John Dizikes, Gary Owens, Edwin Barber, and Eileen Tanner carefully read various versions of the manuscript and offered many sound editorial suggestions.

Katie and Sandy Sanford have consistently supported both the inner and outer journeys.

Extensive criticism and enthusiastic encouragement from Janet K. Larson, a longtime friend and muse, guided the manuscript through its final stages. Kristina Hooper Woolsey and Frank Wiley at Apple Computer generously helped my map making. Holly Panich and the staff at Ohio University Press have been supportive and very skillful in bringing the book to its present form.

AND I THANK friends and strangers alike who, upon hearing of this project, have said, "How interesting! I'll look forward to seeing your books." I needed that.

M. J. Mayer
Santa Cruz, California
December 1988

KLONDIKE WOMEN

INTRODUCTION

"Gold! Gold! Gold! Gold!"

IN MIDSUMMER 1897, sensational headlines from San Francisco and Seattle flashed across an economically depressed world like wildfire. "SACKS OF GOLD FROM MINES OF THE CLONDYKE," announced the San Francisco *Chronicle* on July 15, as the steamship *Excelsior* unloaded weather-beaten miners from the wilds of Canada's Northwest Territories. Some literally staggered down the gangplank, their luggage bulging from the weight of gold. Everywhere in North America and western Europe the news ignited eager talk of the fabulous gold. Minds weary of economic depression and adversity were lit with images of wealth and lavish living.

Subsequent newspaper stories told how a few men and women had worked on their own for less than a year to bring out the great treasure. "WOMAN KEEPS HOUSE, PICKS UP $10,000 IN NUGGETS IN SPARE TIME," announced one article about Ethel Bush Berry. She and her husband, Clarence J., had gone north one year earlier, shortly after they married. They started with practically nothing, but now they had returned with gold worth $84,000 or more. In explaining how they did it, C. J. said, "I question seriously whether I would have done so well if it had not been for the excellent advice and aid of my wife. I want to give her all the credit that is due to her, and I can assure you that it is a great deal."[1] Ethel, in turn, praised her husband. She admitted that they had endured hardships and that she wasn't so sure she was willing to make the difficult trip to the Klondike across the snow and ice again. But she also advised women of what to take should they decide to go.[2] And C. J. declared of the strike, "Two million dollars taken from the Klondyke region in less than five months, and a hundred times that amount awaiting those who can handle a pick and shovel tells the story of the most marvelous placer digging the world has ever seen."[3]

In the midst of a depression and with prospects like that, who wouldn't be excited? So in the days of corsets, leg-of-mutton sleeves, bloomers, and long skirts, many women risked whatever they had to go find their own fortune. They braved the rigors of getting to the Klondike and faced the unknown so they too could pick up nuggets of gold.

The Klondike Women

These are the stories told by some of the women who took those risks to join the great Klondike gold rush of 1897–98. They are the women's own stories of how they tried to get to Dawson. To some it may be surprising to learn that there were any women in the stampede—that is, besides dance hall girls and "sporting women," those who exploited their ability to be useful to men in return for a little of their hard-earned gold. But in fact women from all walks of life joined the rush—poor but aspiring immigrants, professional women, socialites, wives, single women, widows, and children.

The children went simply because their parents did, although both Ethel Anderson (Becker),[4] at five years of age, and Edith Feero (Larson), ten years, seem to have been excited about the venture. Some of the married women would not have been on the trail if it hadn't been for their husbands. But from their diaries and letters, it is clear that others, like Mae McKamish Meadows of Santa Cruz, California, and Emily Craig (Romig) of Chicago, were as enthusiastic about their prospects in the Klondike as their partners. And there were married women such as Martha Purdy (Black), Lizzie M. Cheever, Georgia Hacker White, and Mrs. J. T. Willis, who set out from Chicago, Boston, San Francisco, and Tacoma for the North while their husbands stayed home.

A few of the women went out of desperation, when there seemed to be no way to support themselves or their families at home. In the Klondike at least there was the hope of payoff for their individual efforts. A number of the women were simply swept up by the mass enthusiasm for riches. Or some probably welcomed the excuse for release from boring, routine responsibilities. There were also tourists in the group, like Mary E. Hitchcock, widow of a United States admiral, who traveled with Edith Van Buren, niece of a former president. Professional women, such as Flora Shaw, colonial editor and correspondent for the London *Times*, joined the stampede as part of their jobs. Among the single and married women on their own were miners, business women, shopkeepers, cooks, restaurateurs, hotel proprietors, journalists, physicians, nurses, nuns, entertainers, teachers, and scholars studying some aspect of the new territory. Some went prospecting for wealthy miners to marry. There were also those who went just to see what all the Klondike fever was about, for the adventure of it, and to improve their fortunes if they could.

What stampeders experienced depended partly upon when they went and what trail they used. Several routes through the Canadian and Alaskan wilderness were possible, though only two—the Chilkoot (Overland) trail and the Yukon River route—were well established before the rush. Some of these trails were relatively easy, others were disasters. Across

I–1. Men and women on Chilkoot trail heading for the Klondike in the spring of 1897. (U.W., Winter and Pond photog., 328

these trails the stampede went on through the twenty-four-hour daylight of short northern summers, as well as through the mere twilight and silent darkness of winter. Stampeders fought swarms of mosquitoes and black flies in the warm months and blinding blizzards and subzero temperatures the rest of the year.

Women, especially those traveling on their own, had very little encouragement to undertake the journey. Press reports carried headlines such as "NO PLACE FOR WOMEN."[5] In the article accompanying this proclamation, returned Klondiker, James Christie, commented on a plan for nurses to go the gold fields. He advised that, "a cruel, and probably fatal, mistake will be made if the intention to send the women nurses is carried out. Women," he continued, "are utterly unfit to fight the battle out there. People in the East have not the least conception of the hardships that have to be endured by those who succeed in reaching the country, not to speak of the horrors of the trail leading into the country. The route by the way of the mouth of the Yukon is certainly the best, but even that route entails the greatest hardships upon all attempting it. It might be set down at once as impossible for women to get into the Yukon by the passes."[6]

Ethel Bush Berry, who had gone in with her husband in 1896 across the "impossible" Chilkoot Pass, when interviewed upon her return in 1897 clearly distinguished between women traveling on their own and those with men.

> What advice would I give to a woman about going to Alaska? Why, to stay away, of course! It's no place for a woman. I mean for a woman alone: one who goes to make a living or a fortune. Yes, there are women going into the mines alone. There were when we came out; widows and lone women to do whatever they could for the miners, with the hope of getting big pay. It's much better for a man though, if he has his wife along.[7]

When Ethel returned to the Klondike with her husband in the spring of 1898, she was accompanied by her younger sister, Edna "Tot" Bush, who was very eager to make the trip and considered it a great adventure. Ethel's husband, Clarence, had planned to return to the Klondike alone that season, but when he got to Seattle, he found that he didn't really want to go in without Ethel. So C. J. telegrammed, asking Ethel to join him and to bring Tot too. Tot's letters to their family in California are full of lively accounts of their experiences. In one of these Tot told gleefully of some advice they had received on the train to Seattle when she and Ethel were by themselves.

> One old "freshy" came up to us just before we got to Portland—he had been on the train all the way—and says: "I beg your pardon, but I hear you ladies are bound for Alaska." Ethel gave him a stare and answered, "Yes, that's our intentions." Then he commenced. "Why," he says, "I will take pleasure in telling you good-by, for if you are going to take that trip, we wont see you again. My young ladies, you have not an idea of what you have to go through. You cannot imagin what it is, and by your selves too. You can't even dream, (and we kept mum). Don't you think you are awful foolish, and let me tell you, there was forty ladies from Boston who went to Alaska for the soul purpose of catching a rich Klondyker."
>
> . . . We treated him awful cool, and Ethel pretty near tramped my toes, for fear I would tell something. I hope he finds out now who we are, after giving us so much information. The conductor knew who we were and kept smiling all the time.[8]

Despite the discouraging advice, many women, both on their own as well as with men, joined the stampede. Their reactions to their gold rush experiences varied as widely as their motives. And the challenges each one faced depended not only on the trail conditions but upon the woman—who she was as a person and how she approached life. What one viewed as torture, another took up vigorously as an athletic challenge, and yet another thought was good entertainment, like a trip to an amusement park. The descriptions are so diverse that sometimes it is hard to believe they are really talking about the same section of trail. What one woman expected from men as common courtesy and essential protection for women on the trail, another experienced as an affront to her independence. For the self-sufficient woman, patronizing rules about what she could and could not do really did spoil the fun. Some women assumed domestic roles in many of the traveling groups and accepted these duties as fair trade in an equal partnership, while others resented the endless demands that these roles often carry. Still others assumed privileges and let others serve them.

The stampede stories are about people and how they interacted with each other. Though many women welcomed and even depended on the company of others on the trail, some merely accepted companions as a necessary part of wilderness travel. And others really preferred to be on their own. Many women saw, and thoroughly disapproved of, attempts by some to take advantage of others. At the same time, there were females who depended on "suckers" for their livelihood—the trail was no different from the rest of life.

I–2. Actresses ford Dyea River on Chilkoot trail in 1897 with the help of a gallant friend. (U.W., LaRouche photog., 2014)

Women's reactions to the North Country itself varied widely. For many, its rough and ready life style proved disappointing, even disgusting—living without the convenience of baths, trying to make a meal with river-soaked supplies, rain-soaked wood, and no stove. The Klondike held a temporary interest for those who went simply to amuse themselves or to exploit its wealth or its people. But many of the women developed a true love of this wild and ruggedly beautiful country. They made it their home and eagerly devoted themselves to the active, self-sufficient life they enjoyed there.

Stampeding

Gold seekers poured into Dawson by the tens of thousands in 1897–98. The town perches on a narrow ledge of summer bog where the Klondike River empties into the mighty Yukon River, so it is a natural supply and transportation center for the nearby mining operations. When it was founded in 1896, Dawson was in the westernmost part of Canada's Northwest Territories. But because of the rapid development brought on by the gold rush, this section was split off from the rest of the Northwest Territories in August of 1898 and named the Yukon Territory.

Stampeders came to Dawson by way of five major routes—the Edmonton, the Yukon River, the Dyea (or Chilkoot), the Skagway (or White Pass), and the Stikine.[9] The stories told here are from women on these major trails.

Just getting to the Klondike proved more of an ordeal than many men and women expected. Some journeys that were planned to last a few weeks took years—when the weary stampeders finally did arrive in Dawson, their only discovery was that the rush was already over. Routes that were advertised to be the "most direct" proved to be largely unexplored or fraught with nearly insurmountable natural barriers—cascading rapids, minus-fifty-degree temperatures, ice-choked rivers, impenetrable forests. Trails that were passable and even pleasant under some conditions became life-threatening when weather changed suddenly for the worse. With rain or snow, scenic overlooks became slippery, narrow ledges, and glacier-hung valleys were suddenly death traps, scoured by mudslides or avalanches. And being told there are rapids with twenty-foot waves is nothing like actually negotiating around the whirlpools, "holes," and smashing rocks.

Joining the Klondike gold rush and surviving it required imagination, especially for the majority of stampeders who until that time had led relatively sedentary lives. The first decision to leave home meant confronting the unknown. It meant planning for something for which they had no previous experience. Even the desperate situations of many stampeders were at least familiar, with well-evolved, interdependent routines woven through them. Disrupting those settled patterns to head for the unknown Klondike was, for many, a major emotional and practical undertaking. Some faced it with dread, others with the zest of adventure. Some turned themselves over to "guides" who were supposed to know more. And some handled the uncertainty by just not thinking about it and hoping that they would make it through somehow. Today's travelers can still experience these ambivalent attitudes. Preparations and the anticipated changes are both exciting and depressing. The less the traveler can depend upon conventional living conditions while on the trip—safe and available food, easy transportation, comfortable and familiar accommodations, enough money—the more the potential disruption. Most Klondike stampeders had none of these assurances.

Once on the trail, most stampeders were farther into unfamiliar territory—geographically, physiologically, and psychologically. Almost none had been to the Northwest before, and what little information they had was often inaccurate and distorted. Some were unsure of the route. Very few of the stampeders were "outdoors" people who had *any* experience, let

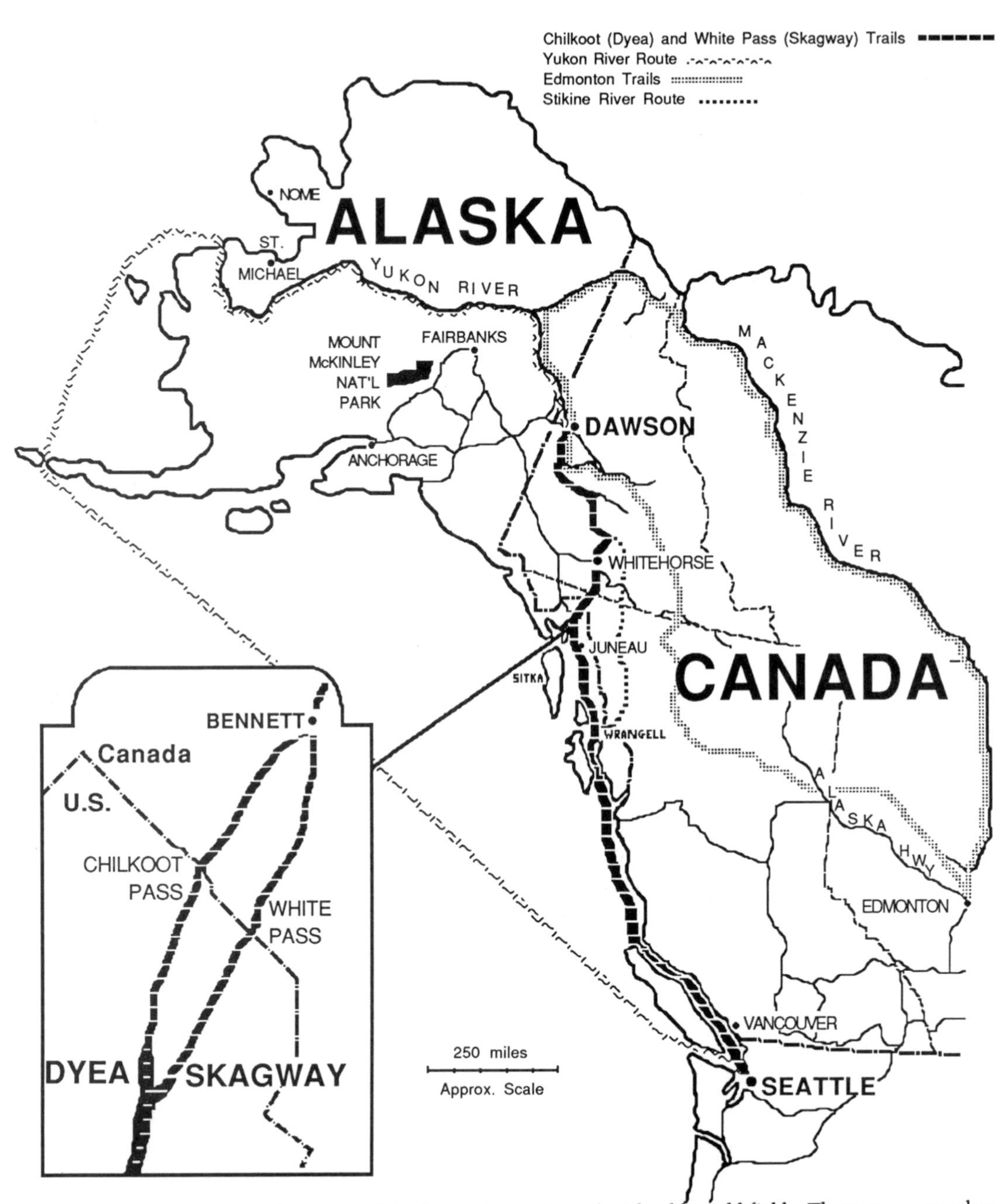

I–3. Modern map showing locations of the five major trails to the Klondike gold fields. There were a number of variations, not shown, of the Edmonton trails. The Dyea/Chilkoot and Skagway/White Pass trails shared all but about 40 miles of the route, shown in detail in the inset. (U.S. National Park Service, modified)

alone developed skills, in packing, tenting, boating, or navigation. Many were not physically conditioned to the strenuous life that most of the trails required. And few had exercised the mettle they would need to see them through their journey safely.

A large percentage of those who set out on the stampede never made it to the Klondike. Of these, a few died along the way. Most simply turned back when they realized the enormity of what they had undertaken.

AT FIRST, stampeders to Dawson did not see the rush as an end in itself. But the stampede did become a central theme of the Klondike gold era because it had real if less concrete value than mining gold and operating gold-related businesses. Gold rush women found value in the human drama of meeting the basic challenges of trail life that suddenly loomed as all-important—staying warm, getting enough to eat, resting, staying alive. They found the courage to recognize their folly and accept failure, to stretch their limits, to face the unknown, to become a hero in a great adventure—or simply to survive. These were what the stampede itself was about. And even for those who did get rich in Dawson, it was often the trail experiences about which they reminisced with most excitement in later years. For most, as in the accounts of women gathered in this book, the "gold" was in the getting there.

WE ARE fortunate that among those who joined the stampede were photographers. Despite the hardships of wilderness travel with the bulky equipment of the day, they captured on film the adventures of the gold seekers, the natural beauty of Alaska and northwestern Canada, and the dramatic transformations of the wilderness that the stampede brought about. We are also fortunate that many women who joined the great stampede to Dawson recorded, from their diverse perspectives, the human drama of those going over the trails. They give a view of the gold rush which is not generally known.

Women stampeders often include details about psychological obstacles on the trails, in addition to descriptions of the physical challenges they faced. Many tell us about domestic life and the concerns of the family, as well as about the business, mining, and political worlds. A number of women were careful observers of their new environment and its native inhabitants. (Their reactions to Native Americans are far from uniform, ranging from curiosity and respect to repulsion.) Though unusually adventuresome perhaps, most were ordinary people and not from privileged families. They protest the injustices they suffer and appreciate the consideration of their companions. The way that they talk and write about their experiences in the gold rush—what they felt they were supposed to do and how they measured up as women—also gives clues to prevailing attitudes about women and the "womanly" in turn-of-the-century America. And sometimes they marvel at what they discover in themselves along the trail.

There is reciprocity, then, in what is revealed by the stories of the Klondike women. Our understanding of a particular time and place in history, the Klondike gold rush of 1897–98, is enriched by the perspectives of women who were making their own ways in this uncertain undertaking. And the gold rush discloses in the women aspects which might have been hidden under more ordinary circumstances.

I–4. Unknown Yukon prospector in 1894 near the as-yet undiscovered Klondike gold fields. (U.W., V. Wilson photog., 1895)

1

GOLD RUSH BEGINNINGS

"Bring me a stocking full of gold!"

IN TACOMA, Washington, in 1979, ninety-two-year-old Edith Feero Larson talked animatedly of her experiences in the Klondike gold rush. As a ten-year-old she had joined the stampede with her family, and she lived the following fifty years of her life in Skagway, Alaska, the gateway to the Yukon and to the Klondike. Edith had been old enough in 1897 to appreciate her gold rush experiences and was well enough in 1979 to recall her early life with detail and humor. She was the last of the Klondike women.

When thoughts of Tacoma, the city of her youth, filled Edith's mind, her cataract-clouded eyes shone with the lively inner images passing before them. Though other ports on the west coast of North America have now become more prominent, Edith proudly emphasized that in 1897 Tacoma was at the heart of outfitting and launchings for the Klondike stampede. To the west, a fine harbor on Puget Sound opens to the riches of the Pacific Ocean. To the southeast looms glacier-covered Mount Rainier (then called Mount Tacoma), surrounded by lush, green forests. Though now silent, it is an imposing reminder of the volcanic eruptions which once formed, and in some places still shape, the Cascade Range.

In 1889, the completion of the cross-continental railroad with Tacoma as its terminus seemed virtually to ensure the new city's prominence in the emerging Pacific Northwest. It was in a strategic position both to reap the harvest of the new state of Washington's wood resources and to be a transportation center. But the worldwide panic of 1892 and 1893, when Edith was five years old, climaxed a long period of economic instability and hit the young city and its residents hard. As Ethel Anderson Becker, another childhood stampeder, later wrote:

> They called the hard times a "panic," but why it came, no one knew. The Northwest should have boomed with the completion of the railroads from east to west in 1889. It did for a few months,

1–1. Edith Feero Larson reminisces as she looks through her collection of old photographs at her home in Tacoma in 1979. She died January 25, 1981, at the age of 93. (U.W., Larson coll'n)

> then money began to disappear and no one had any work. For a while our papa cut firewood for the railway for a dollar a day—a fourteen-hour day. "It keeps us eating," he said.[1]

Hard times continued in the Northwest until the beginning of the Klondike gold rush.

The Feero Family

Emma Babcock Feero and her husband John had come to Tacoma from Maine in 1889, the year Washington became a state.[2] They had four chil-

1–2. Tacoma harbor as it was when Edith Feero Larson was a young girl. Pacific Avenue with its boardwalk is at center, and railroad yards are to the right. The photo was taken on June 20, 1893, looking north from the Northern Pacific Railroad Headquarters Building. (U.W., Waite photog., 98–6)

1–3. Harbor and part of downtown Tacoma are in the foreground with Mount Tacoma (now called Mount Rainier) in the background to the southeast. The photo was taken on December 28, 1894, from City Hall Tower. The *U.S.S. Hasler* is in the left foreground. (U.W., Waite photog., 227–12)

dren at the time—Willie was the oldest, followed by Stewart, then Edith and her twin sister Ethel, who had been born July 17, 1887, in Auburn, Maine.[3] A fifth child, Frank, was born in Washington in 1889.[4]

Edith's father started a transportation business that prospered until the "bust." Edith recalled, "He lost everything but two horses, a wagon, five kids, and his wife. It's all he had left."[5] During these hard times in Tacoma, Edith remembered, "there wasn't anybody didn't have anything. If you didn't have it, you just didn't have it. And we were lots of times hungry. Then we moved from one place to another."[6] Between 1893 and 1897, the Feero's residence changed five times as life got harder and harder. Like many other families, they spent several years just scraping along until news of the Klondike discovery arrived.

1–4. The infant twins, Ethel and Edith Feero in late 1887 or early 1888. (U.W., Larson coll'n)

1–5. The Feero children in about 1890. From left to right are Ethel, Edith, baby Frank, Willie, and Stewart. (U.W., Larson coll'n, Kelly photog.)

1–6. John and Emma Babcock Feero in 1890 or 1891 with baby Frank in a fashionable coat. (U.W., Larson coll'n, Kelly photog.)

The Stampede Begins

In the midst of the bad times, in the wilds of what eventually would be the Yukon Territory of Canada, three Native Americans and a white man found gold. They located it on a creek that flowed into the Klondike River, about two miles above its mouth at the Yukon River. The location was not far from the eastern boundary of Alaskan Territory, which the United States had purchased from Russian in 1867.

Accounts vary as to who made the discovery. Some versions, such as that of Dominion Surveyor William Ogilvie, give credit to Skookum Jim Mason, a Tagish Indian with an impressive reputation for intelligence and strength of both muscle and character.[7] (*Skookum* means "strong" in Chinook trading jargon.)

Others claim it was Jim's sister, Kate Mason Carmack, who made the discovery while collecting firewood or washing some camp pans at the stream.[8] The official, most accepted version of the story gives credit to George Carmack, Kate's Caucasian husband, and to Robert Henderson, another Caucasian prospector, who told Carmack and the Indians of prospects up the Klondike River.[9]

1–7. Skookum Jim on his claim on Bonanza Creek, Klondike District, in about 1897. Photo is by William Ogilvie, who was at the time Dominion Surveyor and later became Commissioner of the Yukon Territory. (U.W., Ogilvie photog., 1898, p. 83, 9072)

In any case, it was a discovery which eventually captured the imagination of tens of thousands of people around the world. "You don't need big money. The gold is there for anyone with a shovel who wants to dig it. You can get enough gold in the Klondike to make you rich for the rest of your life!" they heard.

But at first, only those who were already in the territory learned about the strike of August 17, 1896. Before the winter set in they rushed from the surrounding creeks and towns to stake and begin working their claims. When the weather warmed again in spring of 1897, it was these lucky men and women who left the Yukon with the rewards of their labors—gold from creeks now called Bonanza (where the original discovery had been made) or Eldorado or Gold Bottom. Their arrival in northwestern ports of the United States set off the great stampede, the Klondike gold rush. Edith Feero Larson described what happened.

> There had been lots of gold brought out before, but when the steamer *Portland* came out, they advertised a ton of gold! Well,

1–8. George, Kate, and their daughter Graphie Gracie Carmack are the central figures on the porch of this cabin, which is probably on Bonanza Creek. Skookum Jim, Kate's brother, stands second from right in suit with gold nugget watch chain. The rightmost person may be Tagish Charley, who was also in the discovery party. (U.W., Hegg photog., 859)

> you know, a ton of gold sounds big! That is what started the gold rush—a ton of gold.[10] Before the paper was cold on the print [*sic*], hardly, the rush was started. All that gold! Each one thought they'd go in and grab off a piece. But you know, . . . just as many came out broke, and just as many died in the interior as came out with riches.[11]

The *Portland* landed July 17, 1897, in Seattle. When her precious cargo was weighed, it came to more like *two* tons of gold. Shortly before, the *Excelsior* had arrived in San Francisco with another group of gold-laden Klondike miners. The news went out to a gold-starved world like lightning. Within days, those already on the West Coast were making their way to the Klondike in whatever kind of vessel would float. "The San Francisco *Call*, August 22, 1897, had a list of 31 boats on the way north from the west coast. Now that meant from California clear up. And there were 15,595 passengers on the boats. You know they were crowded."[12]

tricts, which rendered it unnecessary to take furs in with them, and even reduced the amount of woolens to be taken. That being no longer possible, it is necessary for the prospector to supply himself with a complete outfit. We give the following as a conservative estimate of the supplies required for one man for one year:

GROCERIES.

400 lbs. Flour,	3 pkgs. Yeast cakes,
50 lbs. Rice,	6 2-oz. jars ext. beef,
25 lbs. Rolled oats,	5 lbs. Evap. soup vegetables,
50 lbs. D. G. Sugar,	1 qt. Bottle evap. vinegar,
150 lbs. Bacon,	1 pt. Jamaika ginger,
25 lbs. Dry salt pork,	5 lbs. Butter,
100 lbs. Beans,	1 lb. Pepper,
15 lbs. Salt,	1 lb. Mustard,
75 lbs. Dried fruits,	½ lb. each, cinnamon, allspice and ginger,
20 lbs. Coffee,	20 lbs. Candles,
10 lbs. Tea,	2 doz cans Cond. milk,
25 lbs. Evaporated potatoes,	1 tin Matches,
5 lbs. Evaporated onions,	5 lbs. Laundry soap,
25 lbs. Dried beef,	5 lbs. Toilet soap,
8 lbs. Baking powder,	3 lbs. Soda.
1 Commissary box,	

CLOTHING.

1 Suit Mackinaw,	4 prs. Woolen mits,
1 Suit heavy canvas,	1 pr. Oil gloves,
1 Heavy wool overshirt,	1 pr. Rubber gloves,
2 Lighter wool overshirts,	1 pr. High top leather boots,
1 Suit oil skins,	1 pr. Best heavy shoes,
2 Suits heavy wool underwear,	1 pr. Best rubber boots,
2 Suits light underwear, mixed,	1 pr. felt boots,
1 Large silk muffler,	1 pr. Arctic shoes,
1 pr. 10 to 14 lb. blankets,	1 doz. pr. socks, mixed,
1 pr. 8 to 10 lb. blankets,	2 pr. German socks,
1 Broad brimmed hat,	1 Sleeping bag,
1 Sweater,	4 Towels,

3 yds. Mosquito net.

Two pieces waterproof canvas, 6 by 10 feet to cover goods.

Extra lacings for boots, and shoemaker's thread, needles, wax and nails, for repairing.

Pins, safety pins, needles, thread.

HARDWARE, TOOLS, ETC.

1 Yukon sled,
1 pr. Snow shoes,
1 Yukon stove, heavy steel,
2 Fry pans,
1 Gold pan,
1 Nest granite buckets,
3 Granite plates,
2 Granite cups,
1 Dish pan (retinned),
1 Milk pan (retinned),
2 Sets Knives and forks,
2 Spoons,
1 Basting spoon,
1 Coffee Pot,
1 Butcher knife,
1 Can poener,
1 Pocket knife,
1 Hunter's knife,
1 Whet stone,
1 pr. Shears,
1 Miner's candlestick,
1 Emery stone,
1 Axe, single bit,
1 Pick,
1 Shovel, spring point,
1 Broad hatchet, or hunters's axe,
1 Claw hammer,
1 Brace and 3 bits (¼, ½, ⅞-in.),
1 Wipsaw with handles,
1 Hand saw,
1 Wooden Jack plane,
1 Extra axe handle,
6 Hand saw files,
6 8-in. Mill files,
6 10-in. Mill files,
1 2-ft. Rule,
1 Padlock,
1 Tape line,
1 Chalk line,
5 Cakes blue chalk,
1 Compass,
1 Spool copper wire,
1 Spring balance,
25 lbs. Nails, assorted,
1 pr. Gold scales,
1 Money belt,
2 Buck pouches,
2 Hasps and staples,
2 prs. Strap hinges,
5 lbs. Pitch,
3 lbs. Oakum,
3 Balls candle wick,
5 lbs. Quicksilver,
1 Pack strap,
150 ft. ½ in. Manilla rope,
4 pkgs. Hob nails,
1 Draw knife,
3 Chisels, (½, ⅞ and 1-in.),
1 Rip saw,
1 One man saw.

FIREARMS.

1 Rifle, 30-30 Winchester,
Fishing tackle,
1 Single-barrel shotgun,
Ammunition.

DRUGS.

Portable medicine chest, containing selected medicines and drugs.

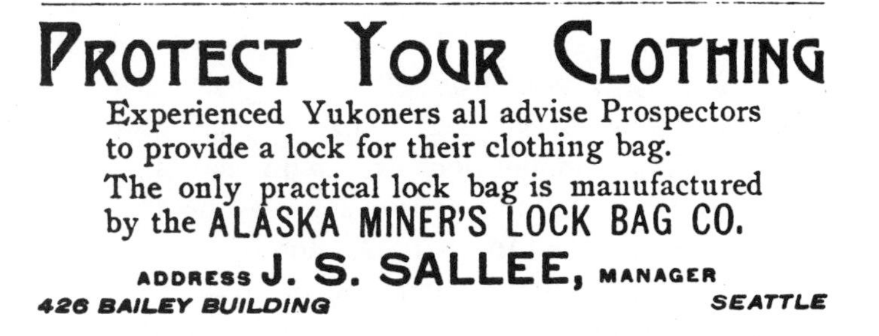

1–9. One list of supplies from a booklet printed in 1897 showing what a typical outfit might contain. (*Official Map Guide, Seattle to Dawson*. Seattle: Humes, Lysons and Sallee, 1897.)

The Klondike gold was an instant sensation. Jurors in San Francisco threatened to quit unless a lengthy trial was completed quickly—they needed to make travel plans. Nuns announced their intentions to head north, along with University football captains who abandoned their teams. Newspapers were filled with interviews of Klondike miners and anyone else who had ever been to Alaska or the Northwest Territories—or even of some who simply had given some thought to it. The press also published maps and lists of supplies that stampeders would need for their outfits.

Long lines of men and women jostled for position at steamship ticket counters, while in jammed corridors outside, would-be gold seekers nearly climbed over each other in hopes of booking passage. Every available boat was soon entirely subscribed for months in advance. At the docks was a frenzy of activity—renovation of ships to squeeze in more passengers, loading of supplies, and people milling about searching for friends about to leave or hoping to catch a glimpse of a real Klondiker. Departures were momentous events, front-page news items, attended by thousands who cheered en masse as each vessel, loaded to capacity, glided from its berth.

Though their bodies may have been uncomfortable aboard those crowded ships, stampeders' minds were transported agreeably enough by fantasies of great riches. Stories of the Klondike discovery left the impression that the gold lay shining in the bottom of streams, "just ready to be grabbed off," as Edith put it. "GOING TO SCOOP UP THE GOLD," announced the San Francisco *Call*, as the *Portland* departed from Seattle on July 22, 1897, for its return voyage north. (In fact, the precious metal was most often in the permafrost, several feet below creek beds, or on surrounding hills in the beds of long-buried rivers from an earlier geologic time.) All knew that dangers lay ahead, but none expected to be seriously deterred by them. All believed they would be getting some of the Klondike gold, one way or another. They just had to get to Dawson.

Expectations of riches were further heightened by the latest gossip exchanged on board—on upper decks stacked with cargo and outfits, or in stuffy, crowded corridors below. By the time she reached Juneau in late August of 1897, for example, Mae McKamish (Mrs. Charley) Meadows wrote enthusiastically to relatives: "We read a letter here from a woman in Dawson, and I tell you it would make you want to have a flying machine and go at once. The people have gold sacked up like wheat lying all around. If we can only get in we will be ready to come back next summer and buy out Santa Cruz."[13]

Among those who also left immediately for the gold country was John Feero.

> Well, Father . . . got a chance to get the help of the man that was later mayor of Tacoma and a man that had a harness shop down on Pacific Avenue. They helped finance [him]. . . . He left mother and us four children here in Tacoma. [The fifth child, Stewart, had been killed by a falling tree.] He took one horse with him, because he couldn't give it away. And he couldn't sell it—it was so darn mean nobody would have it. And it only cost $5 a piece . . . to ship [horses] to Alaska. So [he] shipped it to Alaska, where he got $75 for it when he landed [in Skagway].[14] The man [who] offered him $75 . . . would give him a job for $5 a day and his keep. [He figured that if John] could handle that horse, then he could handle other ones. So he went to work for that man [as a packer on the White Pass Trail] and saved his money so he could send for the family.[15]

In a very short time John Feero wrote to his family to join him. So in August of 1897, Edith and her mother and brothers and sister boarded the steamer *Al-ki*, a freighter converted to handle more passengers, and headed north. Their destination was Skagway, at the very top of Alaska's southeastern panhandle.[16] "When we got onto the boat, everybody was standing, the dock was lined [with people.] Everybody yelled, 'Bring me a stocking full of gold!' They didn't even think of sacks then; they always wanted a stocking full of gold."[17]

The trip north was via the Inside Passage. Stretching one thousand miles from Puget Sound to Skagway, this calm strip of Pacific Ocean lies between the mainland to the east and protective islands to the west. In its southernmost sections, silent, thick, green forests cover shorelines rising steeply from deep, blue waters. Bald eagles survey the waterway from tree-top aeries while crisply patterned black and white killer whales patrol the cold depths. The cool air is usually filled with mist or steady, fine rain, neither of which seemed to dampen stampeders' spirits. Occasionally the calm waters erupt with a Pacific storm. For the overloaded, ill-fitted boats pressed into service for the stampede, this sometimes meant disaster on the rocks. In the northern reaches of the Passage, great, gray, rugged mountains line the route; upon them compressed snow forms glaciers that extend from peaks to waterline. Their blue-green masses of ice groan and grind slowly downward, finally cracking into the water with a retort that echoes for miles. The resulting icebergs form hazardous though infinitely varied floating sculptures in the shipping channels.

Edith's remembrance of her first trip up the Inside Passage was mostly of the people on the boat rather than of the scenery or the weather. Although the *Al-ki* was crowded and far from luxurious, for Edith it was an

1–10. Excitement fills the docks of Tacoma on August 7, 1897, as hopeful stampeders are seen off by envious friends and relatives. At center a crated horse is being loaded on the steamer *Willamette*, much as John Feero's horse might have been. (U.W., Waite photog., 361–10)

adventure. And the ample food, after years of having barely enough, left a strong impression on her ten-year-old mind.

> Well, we were treated pretty good because we were the only children on there. There were some dance hall people on the same boat, and they'd set on the railing on the boat at night and sing such songs as "Hot Time in the Old Town" and "Today's the day we give babies away for a half a pound of tea." (. . . We never dared remember that one. My father wouldn't let us. He said, "You're not going to sing anything like that because they don't give babies away." Babies were precious to him.)[18]
>
> . . . We had a little room with three single berths for five of us. . . . You get one up at a time and put 'em out on the deck, and then the next one . . . you put to bed. There were no conveniences, you know, like today.
>
> . . . We had a kitchen, the little ol' *Al-ki*. The kitchen was . . . up under the . . . bridge on the top deck. All the food had to be

> carried down the deck, into the social hall, down through the social hall, into the dining room. It was a boat that had been crowded in from somewhere for the . . . stampeders. . . . I don't remember how many people were on the boat. There were so many [that] you took turns going down to the dining room. If you had the first table, you'd better be there! They did feed good.[19]

Ethel Anderson (Becker)'s father had also left for the gold fields in 1897. By August of the following year Peter B. Anderson had earned enough money on Eldorado Creek to send for his family. Five-year-old Ethel sailed from Bellingham, Washington, with her mother, Emma, and two younger brothers about a year after the Feero family had left for Skagway. Their six-day trip up the Inside Passage did not go so smoothly, however, as Ethel later related.

> Everyone was seasick—everyone but those Anderson children—and our mama was the sickest of all. One evening Clay, just a year old, cried until the steward came to see what was wrong. Mama lay on the floor, weak from retching. The steward

1–11. Stampeders and well-wishers crowd the dock at Seattle as the *Willamette* prepares to leave with 800 passengers for Alaska. (U.W., Wilse photog., 20)

lifted her gently into the berth beside Clay, then tried to soothe him.

"He won't go to sleep unless he gets a sugar tit," I informed the young man from the upper bunk. "A sugar tit?" "Yes. Can you make a sugar tit? Mama mixes milk and bread and sugar and ties it in a cloth. Clay sucks on it and goes right to sleep."

"I'll see what I can do," said the steward. Soon he returned with the old-fashioned pacifier, and Clay, sucking its moist sweetness, went to sleep. The steward brought Dewey and me apples and told us to be "good children" because our mama was so sick. As if we didn't know.

We ate our apples clear to the cores, picked out the black seeds, and ate them too. Then we invented a game, "Let's see who can make the biggest ball of wool out of the nap of the blankets." No boat ever had such energetic moths![20]

For both the Feero and the Anderson families, heading north to the gold country signaled the end of hard times.

1–12. The *Al-ki* at berth in Skagway, about 1900. Though not a large boat, because of the many years it traveled between Washington and Alaska, it is reported to have carried more passengers on the route than any other ship. (U.W., Hegg photog., 865, Album 16, p. 40)

Gold Rush Widows, Orphans, and Stampeders

The Feeros and Andersons are typical of many families caught up by the gold rush. The husband-father went north alone, then sent for wife and family when a fortune was made or at least living was safe. This is one pattern of how women joined the Klondike gold rush, but it is not the only one.

Most wives never did join their husbands. Instead, they became the links to home that kept many a stampeder true to his purpose. Most stampeders were novices in the wilderness and never planned to spend more time in the North Country than it took to make their fortune. For them the hardships of the trail and the harsh realities of mining and living in the Yukon were endured only so a more comfortable life could be enjoyed back home later.

It was hard to get mail in and out of the Klondike. It traveled by steamer if the Yukon River was open or occasionally by sled in the winter. But the mail was a precious link with the outside,[21] and the correspondence between stampeders and their families back home is one important source of information about the gold rush times. Many wives who stayed at home received no letters, however, either because they and/or their husbands were not literate or because some husbands used "disappearing" in the North as a way to end an unhappy relationship. Of course, some stampeders actually did die along the trail or in the Klondike. Usually their widows would be notified, but some women could only guess at what had happened when reports of avalanches, mud slides, missing expeditions, capsized boats, and forty-below temperatures reached them via newspapers or returning travelers. And yes, there were even "widows" whose husbands suddenly reappeared with bags of gold.

There were also many women who went to the gold rush with their husbands, some with children in tow, others bearing them along the way. Favorite Klondike stories include descriptions of women who worked as partners with their husbands and sometimes of wives more stalwart than their mates. Perhaps more unexpected are the women who went to the Klondike leaving their husbands and families behind.

And then there were the single women, coming thousands of miles to join the stampede entirely on their own.

EDMONTON, THE POOR FOLK'S WAY

"The boys threw dishes at her"

GOLD in the Klondike! But where exactly is that? Many stampeders in the summer 1897 had only a vague notion of what getting to the Yukon involved. To pick a route they simply gathered information from newspapers, pamphlets, handbooks, and gossip as they went. Most made their way to the West Coast because the first boats with all the miners and their gold had arrived in San Francisco and Seattle. That fact gave an immediate boost to these ports, despite the fact that other coastal cities like Victoria and Tacoma were also good places to start the voyage northward.

Stacia T. Barnes (Rickert), a German immigrant, recalled how she and her husband selected their route once on the West Coast.

> In June, 1897, my husband George W. Barnes and I left for Dawson, Y.T., to seek our fortune. When we reached Skagway most everyone was going to Cook Inlet (Sunrise City). So we went with the rest. There was not much Gold there. We expected to leave on the last Boat for Dawson but missed the Boat so thought we would stay with the Camp and try our luck. Spent a fortune looking for a Gold Mine until fall of 1900, when we took the last boat out. Expected to go to Dawson but again we got side tracted in Valdez and Coast Towns.[1]

Eventually, Stacia settled in Fairbanks and became a lifelong resident of Alaska.

"Details" other than the route were also left to chance by many. Tappan Adney, correspondent for *Harper's Weekly* and the London *Chronicle*, reported from Victoria with unconcealed amazement that "two men in three virtually are carried along by the odd man. They are without practical experience; it is pitiful to see them groping like the blind, trying to do this thing or that, having no notion of what it is to plan and to have the ends fit like a dovetail. I asked a Frenchman from Detroit how he meant to get over the pass—was he taking a horse? 'Oh no; there would be some way.' "[2] The Klondike stampede, especially in that first hectic summer, indeed could be a haphazard undertaking.

For many in the East and Midwest, a route through central Canada seemed a more direct approach to the Klondike. Stampeders could travel by rail to Edmonton on the Canadian Pacific line. Then, theoretically, they could go overland along a route approximated today by the Alaskan Highway. The problem was, in 1897 there was no established trail crossing the dense forests, low-lying muskeg swamps, and the majestic continental divide of the Rocky Mountains. Alternately, gold seekers could take boats down the north-flowing Athabasca, Slave, and Mackenzie rivers before going upstream on the Peel and Rat rivers and crossing the relatively low continental divide above the arctic circle into the Yukon River basin. This route was familiar to Caucasian traders and Native Americans of the region. But trappers or prospectors had never used the Mackenzie route routinely to get to Alaska or the most northwestern areas of Canada.

On the Overland Route

Nellie (Mrs. G. E.) Garner of Fresno, California, was one of at least twenty-one women who were among the fifteen hundred stampeders on the various branches of the Edmonton trails.[3] The Garners left Edmonton August 24, 1897, with a group of twenty other stampeders from Fresno. They were, therefore, among the first to search for signs of the overland trail northward. An optimistic Nellie wrote to relatives back home about their warm reception in Edmonton.

> We left Edmonton at 4 o'clock this evening and traveled eight miles, to St. Albert. During all of the time it was raining very hard, and owing to the darkness and disagreeable weather we did not stop to camp, but went to the hotel.
>
> It seemed as though everyone in the town was at the hotel to welcome us, and before leaving Edmonton, by the way, a large crowd of well-wishers surrounded and gave us a sort of "Fourth of July reception." The ladies and gentlemen persisted in approaching and introducing themselves. They all wished us good luck and, of course, gave me plenty of advice as to what I must wear, how to wear it, etc.
>
> A couple of photographers took my pictures in different poses and attire, and a local author is going to write a book about our party. It will be illustrated.[4]

2–1. Modern map showing with double dots the main overland and water routes of the Edmonton trails. (U.S. National Park Service, modified)

2–2. Nellie Garner on horseback and other members of the Fresno party preparing to leave Edmonton in August of 1897. (P.A.A., E. Brown coll'n, B5241)

On their 120 pack animals the Fresno party carried two years' supply of provisions—a ton of goods per person. The Canadian government insisted on adequate preparation by requiring each prospector to bring a year's supplies into the territory. There would be no welfare for ill-equipped gold seekers. Fortunately the Garners had realized before setting off that their trek would be considerably longer than the 700 miles their Fresno-hired expedition leader had led them to believe. He, in the meantime, skipped town with his "advance payments of fees" as soon as they had reached Edmonton.[5] According to Nellie, their first few days on the trail were rather festive.

> Tea parties were planned, and I received any number of invitations to attend them. All along the road and at stopping places people stare at our party, . . . and I often hear them say, 'There goes the woman who is bound for Alaska.' Perhaps I look somewhat odd in my attire as I sit astride my horse with spurs dan-

> gling from my shoes. We are traveling fifteen or sixteen miles each day, and if our present good health continues I think we shall reach our destination about the time we figured on.
>
> We have a guide with us. The kind-hearted people of Edmonton hired him to accompany us. The people of this part of the country make every inducement to those wishing to travel to Alaska by the overland route. They desire them to come this way.[6]

Buoyed by the holiday-like atmosphere that their departure elicited, Nellie's only complaints in this early letter were of frightful weather, of nearly losing one of the horses in a river crossing, and of her annoyance with cooking. According to the newspaper which published portions of her letter, she frequently alluded to her suffering and being nearly blinded by smoke.[7]

Three weeks later the Garners' party had progressed about one hundred miles, and they were struggling to get their horses through miles of fallen timber. A member of the party who turned back on October 2 reported the following.

> We experienced the greatest difficulty in keeping our horses in condition. The feed is good and there is plenty of water, but the difficulties of the trail are death to horses. The trail is filled with spots called 'muskeags,' which are small marshes, covered for a few miles with moss. A horse, stepping on one of these, sinks to his body. It sometimes takes hours to get him out. Then again, the trail was blocked with fallen timber, thrown down by forest fires. It was necessary to jump the horses over this. Such work wore our horses out fast, and we were obliged to abandon many of them along the trail.
>
> We lost so many horses that it soon became evident that we could not all get through. I accordingly sold my remaining horses and outfit to T. J. Kelly, my partner. This will materially aid him in his journey to the diggings.[8]

Of course, such struggling by hundreds of impatient travelers also took its toll on the Canadian wilderness. Many stampeders on the overland trails seemed peculiarly unconcerned that they were, in fact, guests in someone else's land. They left smoldering campfires to burn trees and grasslands. They grazed pack animals in small meadows they found in the heavy forest, leaving little feed for wild animals or Native American's horses. They cut their way through forests and killed wildlife with thoughts only of reaching the Klondike. Some destroyed Indian traps and domestic animals as though they were entitled to eliminate anything that did not serve their purpose. It is not surprising that the stampede was not welcomed by many of the native populations of western Canada.[9]

When winter set in seventy-three days after they left Edmonton, the Garners were at Fort St. John on the Peace River, only 250 air miles from their starting place. By that time the Fresno party had added 94 of their 120 horses to the over 4,000 other animals that would eventually die on the Edmonton overland trails.[10] In the spring, eighteen of the original twenty-two stampeders set out again. They passed through Fort Grahame in mid-June, painstakingly cutting a pack trail through dense forest as they proceeded up the Fox River Valley. By mid-July of 1898 two members of the party were beginning to show signs of scurvy.

In those days it was not known generally that scurvy was caused by vitamin C deficiency, so few stampeders knew how to prevent it. Some theorized that it was caused by a poisoning, allied with ptomaine, from badly preserved fish and meat.[11] Proposed therapy included fresh fruits, vegetables, and meats, and some advocated exercise—something of which the stampeders had plenty. Marie Riedeselle, a masseuse advertising in the June 28, 1898, *Klondike Nugget*, even claimed, "Scurvey prevented and cured by new method. Lost vitality restored." Those who ate only the supplies they carried did not get vitamin C; as a result, many who were on the trails for long periods of time, as on the Edmonton routes, got scurvy. It was, in fact, the Edmonton trails' chief cause of death—claiming thirty-two of seventy who perished.[12] And scurvy is one of the slowest, most painful ways to die. First symptoms are innocent enough—aching muscles and bruised-looking skin. But then one notices that the flesh under the bruise is spongy rather than resilient. Gums become soft and bleed, teeth fall out, and eventually there is hemorrhaging from most mucous tissue. The victim gradually loses all energy, becoming unable to move, and finally welcomes the death that ends long months of suffering. If fresh food is eaten, however, recovery comes within a couple of days.

On July 24, 1898, then, Nellie Garner and members of her party recognized the seriousness of their companions' conditions. Though they were within a few miles of the Yukon River watershed, their progress was nevertheless too slow to continue safely. So Nellie and seven others in the party decided to turn back. She was very disappointed when interviewed by the editor of the Edmonton *Bulletin* on September 5, 1898. It had been over a year since she had set out for the Klondike. The editor wrote, "Mrs. Garner, though realizing that the trip was more arduous than she had ever expected, regrets the circumstances which compelled her husband's and her own return and was anxious to keep on and if necessary to go through to Dawson."[13]

Members of the Fresno party were not the only ones to be disappointed by the Edmonton trails. It was not uncommon to spend a year on the trail, and several parties took as long as two years, for those who survived. One account of this trail reports that "for every single man lost on the White and Chilkoot passes hundreds perished on the Edmonton Trail."[14]

A more systematic study shows that the outcome was not so drastic. Of the approximately 1,500 people who set out from Edmonton, over 700 arrived at Dawson by 1899. Of the 775 who went overland, 160 reached Dawson and 35 are known to have died. Of the 785 who went by the water route, 565 got to Dawson and another 35 died along the way. The others either turned back before it was too late, or became settlers somewhere along the routes. However, this systematic study probably underestimates the number who died because it counts only confirmed deaths on the trail and does not include those who reached Dawson but died soon after from effects of their strenuous journey.[15]

For example, what happened to the Larrabees? Mr. and Mrs. G. W. Larrabee from Buffalo, New York, were among those who set out from Edmonton in the early months of 1898 to travel across the snow to Dawson. With them was their three-month-old daughter. Very little is known of what happened to them, but it is apparently their child whose tiny grave was found about six miles from Fort Assiniboine, about 90 miles from Edmonton.[16]

Reports from the Water Routes

Though the water route of the Edmonton trail was considerably longer than the overland branch, it was nevertheless the better. Among those who took the water route were Emily Craig, her husband A. C., and their party.

Emily Craig was a Danish emigrant, living in Chicago with her husband when news of the gold discovery in the Klondike infected the city with gold fever. The Craigs were stricken along with the rest. They joined a party of eleven others under the direction of Lambertus Warmolts. Emily apparently kept a detailed diary during her trip, which several decades later she elaborated to describe the beginnings of her life as *A Pioneer Woman in Alaska*. (After the Klondike gold rush she became a longtime resident of Alaska.) It is obvious that Emily saw herself as a full-fledged partner in the stampede party. And she showed impressive sensibility and ingenuity in the midst of frustration as she adapted her jobs—primarily cooking, camp maintenance, and clothing fabrication and repair—to difficult trail conditions.

Warmolts, the expedition manager, like the leader of the Fresno party, claimed to know the Northwest well. (In fact, he had never been there.) He confidently promised to have his party to the Klondike in six weeks. (In fact, it took Emily and A. C. Craig two years, and many in their party never made it.) His fee was $500 per person, for which Warmolts agreed to supply guidance, travel, and the requisite food for a year. On August 25, 1897, encouraged by cheering well-wishers, the stampeders left Chicago by train.

One of their first concerns was food—what to take and how much of it. Lists of suggested supplies had been published in many newspapers and pamphlets, but the lists varied considerably in completeness and in recommended amounts. How much to take, of course, was influenced by how long one expected to be on the trail. And for those provisioning en route, needed supplies were often not available. Emily was unofficially designated as cook-without-pay, a job which she grew to hate because of its unending demands. And as she explains, their provisions proved inadequate in quantity and variety. So it was hard for her to appreciate the advances of technology which offered, for example, commercially dehydrated foods, when she had to use them under such adverse circumstances.

> When the party arrived in Calgary, Canada, our manager bought our provisions from the Hudson's Bay Company—4400 pounds of flour, 500 pounds of corn meal, 1300 pounds of dry beans, 2700 pounds of salt pork, tea, coffee, dried apples, matches, and numerous small articles. There were very few things in the market as other parties had outfitted before us. Fruit and vegetables we could not have carried, and canned goods were scarce. There was little milk, and it was Borden's, thick, sweetened milk. Saccharine was procured instead of sugar. That was a mistake, as no one liked it. We did get dried potatoes, dried onions, and dried eggs—no fresh or canned meats. The variety of our bill of fare was thus very limited, and for one who had never cooked, except for my husband and myself, cooking for fourteen, with no grocery store handy, was enough to give one gray hair. This kind of diet was new to me and new to those who had to eat it. Everyone missed butter, but salt pork grease was used instead, and it was not so bad after one became accustomed to it.[17]

The Craigs departed from Athabasca Landing, ninety miles north of Edmonton, in their newly christened boat, the *Emilie*. Emily's description hints of the trouble that lay ahead.

> After the boat was loaded, we pushed off downstream, with the advice from the trader not to run the rapids alone, but to secure Indian pilots to do this for us. Our oarsmen had never rowed a boat, and our pilot or guide had never handled a sweep or steering oar, nor did he know how to handle a square sail. He could not tell us how far ahead the rapids were, nor when or where we would get Indian pilots. The faces of each of the party bespoke disgust for the guide, and doubt as to his ability.[18]

Emily's first encounter with Native Americans was with those who were to serve as their pilots through the rapids. She obviously found them interesting and was really perturbed when one member of her own party, a ventriloquist, insisted on practicing his art with the natives. (He had already lost points with Emily by making "the stove talk, to the disgust of the cook—and that was me."[19]) Emily was eager to learn from these local experts the ways of the wilderness she was entering, and her book contains many interested, if not completely sympathetic, observations about their practices.

> Some miles down the river we saw five Indian tepees, or summer tents, in a little cove. They were made of poles tied together at the top and covered with birch bark and moose hides. On the beach were some very good birch-bark canoes, and on the bank four or five well-built Indian men. They had prominent noses and long hair to the shoulders, and wore moccasins. The women and children were dressed in bright print clothing, and all wore moccasins. In a smokehouse was some moose meat smoking, and on a cache in the trees was plenty of moose meat, as they had killed three moose. This was the camp of Chief Big Stone, the pilots who were taking all of the miners through the thirteen rapids in the next eighty-five miles. We secured a quarter moose meat for ten cents a pound.
>
> . . . The Indians, or pilots, were very polite, and removed their hats when they ate, a thing some of our boys did not do. They ate outside the tent, as they did not like the sound of a peculiar voice, or the sayings of the ventriloquist—I felt like killing him. . . .
>
> I was very interested in the tepee and the way they camped, as it was certain we would have many camps ahead, and I wanted to learn as much as possible about camping. The Indians are very kind to their children, and were pleased to talk about them. However, I was told that when they get too many girl babies, they just throw such a baby out in the snow and let it freeze, as girls are not so useful and desirable as boys. If they were stringing me along with a false story, I hope they drown sometime themselves, as they made me shudder. These stories ended when all the party was together again, and we started for Grand Rapids.[20]

Going through the various rapids was a tiresome process of lightening the boat's load above, packing it around, running the boat through, then reloading at the bottom of the rapids—the whole sequence to be repeated at the next set of cascades. If the boat ran into problems in the rapids, even more work was in store.

> . . . our boat was standing on a rock in the middle of the rapids and on each side roaring, running water was going about twenty-five miles per hour. The boys that had walked ahead of us had to go to the other boats and ask them to come back and help. They had to unload their boats first, then track or pull them upstream with a rope from the bank against this swift water for a distance of about a mile, and then with a long rope allow their boat to swing out to our boat before the provisions could be transferred. This was dangerous business, and plenty of work for all hands. Mr. Springer was stationed on the bank to keep up a fire so as to guide the men in the boats.[21]

Finally, at about midnight, the lightened *Emilie* swung loose from the rocks and was brought to shore with two holes in her hull. Emily wrote, "About 1:00 a.m. we rolled into our sleeping bags, and we did not need to be rocked to sleep, either."[22]

2–3. Boat runs one of the many rapids on the Edmonton water route, this particular one on the Slave River. (P.A.A., E. Brown coll'n, Mathers photog., B2880)

Even when all went well, the portage of supplies around rapids was arduous. By the end of the day, there was little energy left for anything but eating and sleeping. Fortunately for the Craig party, Emily made sure that good meals were ready whenever the "boys" were. But as time went on, the resources of many of the group members became more and more depleted.

> Four of the boys who were walking had not arrived by nine o'clock, so Mr. Springer took a lantern and started back along the trail to look for them. He had the light behind him so it would not blind him on the trail, and so did not see Graham and Mr. Card coming toward him. Mr. Card, thinking he would have some fun, fired his revolver and scared poor old Mr. Springer so badly that he never liked Mr. Card after that. Mr. Card and Mr. Graham came on into camp, but Mr. Springer went on up the trail to look for Mr. Hore and Mr. Carter, who was sick and had a lame foot. Mr. Springer came back about midnight, having missed them, but a little later they came in, all tired out, and Mr. Carter a very sick man. He had wanted several times to lie down and die but Mr. Hore would not let him. I wondered if he would last the trip.[23]

At Grand Rapids, the strain of the trip on Emily began to show. A tramcar had been built at the rapids to help portage supplies.

> . . . We got up early and had our first boatload at the tramcar landing at 9:00 a.m., and by the following morning all our goods were over the tram route and ready for the boat. That day I would have returned to Chicago had it been possible. The boys thought, besides doing the cooking, I should help pack some of the things to the tram. I got pretty peeved, and while Mr. Craig was gone with a load, I started for the tram with some of the lighter things, but when I met some of our men they made me return to camp. Mr. Craig also was pretty angry when he heard about it. At times now I can see it would have been better if I had stayed at home.[24]

But Emily hadn't stayed home, and she didn't turn back. Past the rapids the water was relatively safe, but the work was no lighter. The low autumn water meant that the boat was constantly being hung up on sandbars and often was released only by great effort. Other stampeders they met along the way were faring no better, and several were already retracing their steps to go by way of one of the coast routes rather than continue.

By October 4 the Craigs had reached the northern boundary of Alberta, but winter was already in the air.

2–4. Portaging a scow around Smith Rapids on Slave River, using logs as rollers. (P.A.A., E. Brown coll'n, Mathers photog., B2910)

2–5. Scows, towed by steamer, approach Fort Resolution the summer of 1901. (P.A.A., E. Brown coll'n, Mathers photog., B2937)

> . . . We got up shivering with cold as it was freezing very hard. The cold came suddenly, and we felt it, as it had been damp most of the time. The Indians started to load the oxcarts, but Mr. Craig found we had to have three more carts, and was forced to hire that many more, making ten in all. The wagons had wooden axles, and the harness was of old rope. The trip began; we tramped through mud and water up to our knees, and the oxen sank in up to their bellies. When the oxen got stuck, the Indians would cut down a small tree and beat them with it, and when I begged them not to be so cruel they laughed at me. The harness would break and then the oxen would run away, and it would take at least a quarter of an hour to find them and get them back to work.[25]

The Craig party arrived at Fort Resolution at the entrance of the Slave River to Great Slave Lake on October 12. Now their predicament became clearer.

> We could not get a guide to take us across Great Slave Lake, as the Indians said it was too late. We were told we could winter here, or wait until the lake froze up and cross by dog sled, as the lake once frozen does not open up until late in June. While we were eating, the wind changed and the guide told the men to pull

up the boat or we would be liable to have no boat in the morning, for the ice would crush it. Mr. Warmolts had had many chewing matches with the men, and we could see it was impossible to get to the Klondike this year. The food would be short, and he had all the money. It looked clearly to all that we would be cheated out of our transportation and money, too, and there were some pretty angry men in camp. He and Ed Charlton were to keep watch that night, but went to sleep, and the dogs ate up twenty-five or thirty pounds of provisions for us. After working so hard to bring the supplies, and knowing that every pound would be valuable this winter, you may be sure this made the men mad, too. They just wanted to be mad about something anyhow.

. . . On October 14 we decided not to have more than two meals a day, to save food.[26]

Within a few days Warmolts deserted taking all the remaining cash. So the party decided to split up. Although breaking up was painful, Emily had some reasons for feeling relieved as well.

In the division of things some of the men were so mean and jealous that they wanted to cut the boat in three pieces, one third for each party, but finally common sense prevailed and they sold the boat to the Hudson's Bay Company and each took an order for his share, to be paid in goods at the next store. They even wanted to cut the stove in three parts, but finally let Mr. Craig and party have it for some of our provisions. When the division was complete, our party moved in next door. Our place was the company's carpenter shop. We paid fifty cents per month rent. It was much easier to cook for four than for fourteen.[27]

The fact that credit for the goods exchanged for the boat was to be redeemed at the next Hudson's Bay store rather than at Fort Resolution is an indication of just how strained the resources at the fort were that winter of 1897–98.

The Craigs spent the winter at Fort Resolution getting to know the post's occupants, learning about the wilderness, and building up their meager supplies to be ready for the spring breakup of ice. Here A. C. Craig's skills as a carpenter were brought into play as he built sleighs, dog sleds, and ice boats for himself and others. Emily kept camp in the Hudson's Bay Company's carpenter shop. She spent much of her spare time with Miss Gaudett, who was part Indian and whose brother ran the company store. They took long walks, talked of Indian traditions and ways, and Miss Gaudett showed Emily how to sew and embroider moose-skin garments.

2–6. Native American skin lodges around Fort Resolution. (P.A.A., E. Brown coll'n, Mathers photog., B3081)

In December part of the original party tried to set out by sleigh, but with discouraging results. Their misadventures illustrate how inexperienced in the wilderness many stampeders were.

> These boys had two sleighs with six hundred pounds to the load for two men each. They got tired when about three miles from the fort, and decided to come back. It took them seventeen hours to make the trip out and back. Joe Piersoni refused to go any further, and just laid down on the lake to die, he said. They were back at about four o'clock in the morning, all tired out from pulling the sleigh. They left one sleigh out on the lake, and Mr. Gaudett [the Hudson's Bay Company representative] had to send out for it. They were in a rather queer position, as they had sent part of their goods on ahead, and sold part of their bacon for sixteen cents when it was thirty-five cents a pound here.

> They had some queer experiences in the short time they were out on the lake. They had two loaves of bread, and when they wanted some of it, they found it was frozen. They tried to cut it but could not; then they used the ax on it, but could only dent it, so they got angry and put it away. Towards evening they went ashore and started a fire, but soon the snow melted and the fire disappeared in the melted snow.[28]

By February Mr. Craig and his partner Ed Charlton began sledding provisions across Great Slave Lake for others to make some money. On April 19, 1898, the Craigs were ready themselves to leave. The ice on the lake was beginning to thin, so it was important for them to get across while they could still use the sleds. Emily emphasizes the difficulty of loading the sleighs, which were too small to carry everything; hard decisions had to be made about what to leave behind. No doubt this contributed to the tension she reports at this juncture.

> By this time all the members of our original party had split up and joined other parties. Quarreling and split-ups was the order of the day. From here on my husband and I remained in our own party, and only for a short time did we have anyone traveling with us. I was so glad to get away from the cooking for a lot of men and their grumbling when food was scarce. We were happy once more to be alone and free—alone where nature's scenery was grand and the very solitude drew us closer together.[29]

For crossing the fast-thawing Great Slave Lake, Emily and A. C. joined with Ed Charlton, from their original group, and two men from another party, Charley and Arthur Griffin. The Griffins had left from Edmonton at about the same time as the Craig's party and had also been forced to winter at Fort Resolution. Charlton and the Griffins would be the Craig's occasional companions as they all worked their way northward from Fort Resolution.

> The ice was bad, and the dog's feet were sore. It was hard to say good-by to our friends here, because they had been so good to us. That day we made Dead Man's Point, and early in the morning we were off again and reached Half Way Point at 5:00 p.m. We then traveled six miles, and as it was hot on the ice, we stopped until 10:00 p.m. We started for Point Press to Hay River, a distance of twenty miles. About halfway across this bay, Mr. Griffin played out and wanted to camp, but I objected, because I was cold and the wind was blowing. We did stop and have a cup of tea, where Mr. Craig had cached some wood. This was his twelfth round trip, and he had traveled over twelve hundred miles freighting and knew where to stop for tea. I had walked a lot in the wa-

> ter, and when I was too far behind would run to catch up. I, too, was cold and exhausted and just could not run or walk any more. After a two-hour rest, we started again. I was tied on top of the load, because I was tired, wet, cold and sleepy, and completely exhausted. I could hear the splash, splash of poor Mr. Craig's tired feet as he ran, and at times helped the dogs, and the swish, swish of the lame and weary dogs' feet as they trotted along in the slush and water.[30]

It was the beginning of May before the ice went out. In the meantime, the Craigs began to build their boat for the Mackenzie River portion of their trip. As on other Klondike trails, lumber for the boat was handsawn, and Emily's description of the process of whipsawing gives an idea of why friendships and partnerships were often cut along with the wood.

> We have several logs hauled out and up to the saw rack. Mr. Craig hewed them flat on one side, and on the morrow we started the sawing. I was the under sawman. The logs were all put on a frame higher than a man's head. A line was made above and below, by chalking a string, raising it and letting it snap back, then the work began. The one above, and that was Mr. Craig, lifted the [whipsaw] up. The one below pulled it down, and that was me. You looked to see the line to keep the saw straight, and got sawdust in your eyes, and as you stooped, some more sawdust went down your neck. It took us several days to cut our lumber. Nails were scarce and all we could get out of boxes were saved. Many wooden pegs were driven to take the place of nails. I held boards, made pegs, and at last went into the spruce timber to gather resin for sealing up holes or cracks. The resin was placed over the crack, and the ends of two burning sticks approximated on either side, and the resin was run into the defect like solder is placed on tin.[31]

After eight days of labor, the boat was ready, and the Craigs set out on the next leg of their journey on May 19. Emily worked along with her husband and Ed Charlton to pilot the boat, lift it off of sandbars, and guide it around obstacles.

> We went ashore to eat our supper, and by the time we had finished, a heavy gale sprang up, and we had barely time to unload our boat before it was full of water. The water kept rising and we had to move our goods three times. When we were taking our boat around on the lee side of the island, Mr. Craig tracking and Ed and I leading the boat, some way I got on the outside and nearly drowned in a deep hole which was full of ice. I did not let go my hold on the boat, but as it was, got in the ice and water up

> to my neck. Mr. Arthur Griffin had just had a similar experience, and when he saw me out there in the water, he got so scared he could not speak but ran and pulled the boat in. It was dreadfully cold and we were glad to get a fire started. It was rather chilly to strip and change outside with a strong wind blowing. Neither Mr. Griffin nor I felt the worse for our cold bath. While I was getting supper, Mr. Craig set his net and caught four fish.[32]

But an icy bath was not the worst of their troubles.

> May 25 was a day I will never forget, for when the only home you have is a tent, and that burns down on you, it makes you feel rather bad, especially when there was not a chance to buy another one. Mr. Craig and I had just walked a little way and the Griffins and Ed, camped near, did not notice the fire until the poles fell in. They did save a lot of powder and ammunition from the tent. As it was, Arthur Griffin went around with a big hole in the back of his coat which burned there while he was trying to save some of our things. I felt badly as it was my fault, for I was baking bread and left a good fire in the stove, the sparks from which had caused the trouble.[33]

Like many others on the water route, the Craigs succumbed to the temptation to take a "shortcut" across the continental divide at a stream entering from the west. They lost seventeen days on that detour up the Root River. They returned to the Mackenzie River but never did hear what happened to the others who continued up the Root. Finally, on July 8 the Craigs arrived at the Peel River and began pulling their boat up to Fort McPherson.

> About five o'clock we reached Peel River, and I celebrated the occasion by falling overboard and getting good and wet.
>
> July 9. We started upriver, and now is where the fun began. Tracking, if you call that fun, is the hardest fun I have ever known. We had to cross the river so often the banks being so steep and so many old trees you could not walk. We reached Fort McPherson about midnight. They just turn night into day, as they have the midnight sun here, and as it is a little cooler during the night they sleep in the daytime and work at night. The report was not very encouraging. Some had gone to Peel River, thinking they might find their way across, but as no one had ever gone that way, and the route was uncertain, some had taken the overland trail, packing from thirty-five to fifty pounds on a dog. One party had eighteen dogs, and the poor brutes were nearly blind from black flies that settled in their eyes, ears, and nose. Others had gone by

2–7. Fort McPherson in about 1900. The Hudson's Bay Company "forts" were often little more than a small group of buildings around a trading post and storehouse. (P.A.A., E. Brown coll'n, Mathers photog., B2868)

> way of Rat River, but the water was so low now they could only take small boats. Some returned from Rat River and were greatly discouraged.[34]

Emily's firsthand account of the realities at Fort McPherson contrast sharply with one early report that misled so many people. Appealing to international rivalry, it advised, "Any Canadians who are anxious to get into the Klondike ahead of the Americans can leave between now and August 1, reach Fort Macpherson, and if winter comes on they can exchange their canoes for dog trains, and reach the Klondyke without half the difficulty that would be experienced on the Alaska route." Even more disastrous was the information that "travellers need not carry any more food than will take them from one Hudson Bay post to the next. . . . You don't need a couple of thousand dollars to start for Klondyke to-morrow by the Edmonton route. All you need is a good constitution, some experience in boating and camping, and about $150."[35] It was because of such early publicity that the Edmonton trail became known as the "poor man's route." But as the Craigs had already discovered, few were willing to risk losing their valuable dogs on an uncertain journey to Dawson, and many of the

Hudson's Bay posts marked on maps were no longer in use. Those that did exist were not prepared to supply hundreds of new people and their dogs with food. And even if dogs and food had been available along the way, they would have cost much more than $150.

At Fort McPherson, Emily met Mrs. Stringer, wife of the local Church of England minister. Soon another female stampeder, Mrs. Braund, arrived. Her attitudes and aptitudes contrasted sharply with Emily's; still Emily could be sympathetic to Mrs. Braund's predicament.

> July 14. There was much excitement—a steamer was coming, and it turned out to be the the *Enterprise*. A party from Detroit was on her. Mrs. Stringer and I went down to see the novelty. We heard there was a lady on board, but that she was too tired to come ashore. The next day at eleven o'clock the lady, a Mrs. Braund, came ashore and had tea with us. I did not like her husband very well, but she seemed like a nice lady and did not like this trip at all. The following day we two ladies discussed our trip and the experiences. I thought she would be better off at home. She told a hard story—started cooking, but the boys did not like her cooking and threw dishes at her, so she quit cooking. She did not like the natives and the handshaking, and the dirt in general.[36]

The Craigs traveled with the Braunds westward from Fort McPherson up the Rat River to Destruction City, so named as the spot where boats were torn apart, the river no longer being deep enough to float them. They could then be used to build temporary shelter. Or the pieces could be portaged a mile or so across the continental divide to the headwaters of the Porcupine River and reassembled for the 330-mile downstream trip to Fort Yukon on the Yukon River. The 400 miles from there upstream to Dawson could then be by way of Yukon steamer if the timing were right.

> July 25. I fared well for my twenty-seventh birthday. I received one package of dried cabbage, one package of dry soup tablets, and a loaf of graham bread. We sailed at 6:00 p.m., and met two men returning with a discouraging report about Rat River, but we decided to see for ourselves. . . . Mrs. Braund's little dog fell overboard and Mr. Braund got a black fly in his nose. He sneezed so hard his false teeth flew overboard. The next night the dogs and black flies kept us awake all night. The following day Mrs. Braund and I tried to have a bath in the river. We saw something like a dog but it turned out to be a lynx. There were parties coming and going all the time these days, and most of them were traveling faster than we could go. I steered today and the boys tracked.[37]

The Braunds, after several days of quarreling, returned to Fort McPherson.

At Destruction City the Craigs again ran out of money and provisions, so they decided to winter there and earn supplies by hauling for others. They built a little cabin which though small was still an improvement over the makeshift tent they had been using. Many others spent the winter of 1898–99 in the vicinity of McDougall Pass as well. They planned to pack across on the snow and to catch the first water of the spring in the Porcupine River on the other side. Most hadn't the practical lessons of survival that the Native Americans had shared with Emily, so many died of scurvy and freezing in the isolated camp, well north of the Arctic Circle.

> Oct. 3. . . . My last winter's experience comes in good now, as Miss Gaudett, who is part Indian, taught me how to prepare fish and game so as not to lose its flavor or its goodness as food. Moose and caribou are tender meat, and she taught me to fry it in a hot pan, turn it often, and never to cook all the red out of the meat, and to pour the juice over the meat when done. Fish was to be fried quickly but not hard, and steamed a few moments and the juice to be poured over the fish, as much of the flavor is lost if the juices are lost or thrown away. The meat and fish should be a little rare and that would prevent scurvy and other sickness. How true we found this later in the winter, when many of the miners had scurvy from eating foods they had along and not getting fresh meats. Mr. Craig was always getting fresh fish and meat for us and for the dogs. Some of the men were too lazy to fish or hunt, and they paid dearly for that kind of life. The men who had native wives had good reason to brag on them, for they could go ahead on snowshoes, if the trail was bad, or could take the handle bars and manage the sleds as well as a man; were good at making camp, kept the moccasins and mittens in repair, and cooked as much as they knew how, and that was game and fish. Scarcely ever did a squawman get scurvy, and that speaks well for the squaws.[38]

A highlight of their winter was a New Year's trip by dog sleigh to Fort McPherson, where Emily arrived in time to help Mrs. Braund with the birth of her baby boy. The Craigs spent twelve days visiting, helping the Braunds and picking up supplies before a letter arrived urging them to return to Destruction City. Many were sick and dying there, and Emily had become the unofficial nurse for the camp. On the return trip, in almost unbearable cold, they reached an unoccupied cabin where they stopped to make a fire and get out of the weather for the night.

> . . . Corby [another stampeder whom the Craigs had met on the trail] started a fire in the fireplace. They put too much wood on, and before we knew it the place was on fire and there was nothing we could do to put the fire out, as we had only a frying pan to shovel snow with, and so the place burned down. They did make a cup of tea on the coals for we had not had any food since four o'clock in the morning. There was nothing for us to do but hitch up the dogs and start for home. On the way it was dusk and hard for me to keep the trail when I had to run. I fell in the snow very often and the dogs would get a good start before I could get up again. I had to go back to the sled and ride, and I was very cold. We reached home before supper, tired and cold and very hungry, having traveled in a temperature of fifty-five below all day. We slept well that night. For the next few days I spent the time cleaning the cabin, washing clothes, and cooking for ourselves and the sick, while the thermometer stayed at about forty-six below.[39]

By April the long winter was over, and on April 18 the Craigs set out again for the Klondike.

> I must have walked twelve miles, but the weather was so nice and the scenery so beautiful, and I was so happy to be on the way, that I jumped on and off the sled if the going was good or bad. I felt like a happy little girl on this nice day to be released from this dreary camp and actually going again. The Rat River was crooked, and the trees hung over the bank; the scenery seemed better at every turn. Going down hills I would hang on the sled and have a regular toboggan ride. Happy childhood came back to me, and I had many a good laugh.[40]

When across McDougall Pass the Craigs once more had to whipsaw lumber and build a boat, but they were ready when the ice went out May 31. By June 11, they arrived at Fort Yukon on the Yukon River. Here they had to stop again to earn money—this time for their steamer passage to Dawson. Emily sold bread, and A. C. did carpentry and odd jobs. The Braunds also made it to Fort Yukon, but went by steamer down the Yukon River rather than to Dawson. Reports of travelers from Dawson were gloomy, so perhaps they had decided to join the new rush to Nome, Alaska, at the Yukon's mouth. Or perhaps they had simply had enough and decided to return home with their infant while all were still alive.

Not until August did the Craigs have enough money to buy passage to Dawson. Their penniless situation is made all the more poignant by Emily's modest concern to feel presentable after two years in the wilderness.

2–8. Emily Craig (Romig) in about 1940 with the flag that she made of flour sacks and red calico at Destruction City in the winter of 1898–99. (Romig, 1948)

. . . I felt as though my clothing was far out of date, and worn out too, and as we were going to Dawson soon, I wanted to look a little better. It was a sight to see the dresses and hats that the ladies, going through here, were wearing.

After our tent had burned, and I had lost all my dresses except the one I was wearing, I was desperately in need of clothing. I secured a little calico and made another dress, and from the Indians we bought a buckskin dress. It was beaded and had fringes at the top of the shoulders and around the bottom. The trimmings were of red flannel and beaded. I also got a fur cap and buckskin moccasins. This outfit was warm and would not tear when we were in the woods. It was good for the trail, and at times Mr. Craig would call me his little white squaw. That was a comfortable dress for the wilderness, but I did not feel right in it when we reached the Yukon and met many whites. The hat I bought [from a woman at Fort Yukon] did not blend with my buckskin suit, and my fur cap was no ornament, even if warm. There was no white women's clothing in these parts, and it seemed everyone was looking at me. I wanted to cry, but that would not make it any better. These were all I had, and I wanted some new dresses before I got to Dawson.[41]

On August 30, 1899, the Craigs at last reached Dawson, two years after they had set out thinking it was to be only a matter of six weeks. As everyone had been reporting, the Klondike rush was over.

At least the Craigs had survived the Edmonton Trail. The Hoffmans were not so fortunate. They were both recent German immigrants to Sandon, British Columbia, and Mrs. Hoffman was pregnant. With them was a German girl, Rosie, whose last name and age are unrecorded.[42] They left Athabasca Landing when the ice broke in late April of 1898. When they reached Great Slave Lake in mid-July and began the 170-mile crossing of its unpredicatable waters, they met with disaster. "He was standing at the sweep, and she was bailing out water, and when she looked up there was no one at the sweep; the other members of the party were at the oars. She asked, 'Where is Frank?' and no one answered. She looked out just in time to see his head disappear. They drifted around for two days, but could not find his body. They had been married two years."[43]

The above account is from Emily Craig, who met Mrs. Hoffman October 3, 1898, at Destruction City. Emily says, "The story of this poor woman affected me so that I was sick for two days."[44] Now what was Mrs. Hoffman to do? A. D. Stewart, a fellow stampeder, wrote the following description of their predicament shortly after Frank's drowning. "The wretched widow is in a most deplorable condition. She has lost nearly all her provisions and most of her clothing; her husband and protector is dead; she is thousands of miles from her German home; and her coming confinement stares her in the face. What will become of her? What can be done for her? God pity and help the poor creature!"[45]

Ironically, Mrs. Hoffman made it to the Yukon; A. D. Stewart, who wrote the above did not.[46] Mrs. Hoffman undoubtedly was aided by her fellow travelers. But then so were A. D. Stewart and many others of the unsuccessful. Perhaps Mrs. Hoffman was spurred on by the single-minded energy that often accompanies pregnancy, keeping the mother focused to take care of herself and her infant. She joined another group and continued. Mrs. Hoffman apparently gave birth, but there is no record of it. The next spring she went on to Dawson, with 170 others who had survived the winter in the Fort McPherson area.[47]

IF ALL the trails had been as difficult as the Edmonton routes, the Klondike stampede would have been more a loping canter. But there were shorter, more reliable ways to reach the Yukon.

STEAMING IN VIA THE ALL-WATER YUKON ROUTE

"I came 13,000 miles without a cent, so I'm satisfied"

If the Edmonton route by way of the Mackenzie River was the "back door to Dawson City," then the all-water Yukon River route was the "side door." As steamers followed the Yukon upstream, they were led northeastward across nearly the full width of Alaskan territory before turning southeast at Fort Yukon and heading into Canadian territory.

In contrast with the Edmonton trails, the Yukon River passage was very well known. Since there were no railroads into the wilderness, the river was the way supplies had been brought into the interior by Alaskan trading companies for many years. And unlike the Edmonton trails, the Yukon River had the reputation of being the safest way in, as well as requiring the least effort. But it was open only for the two to three summer months when the river was free of ice but still had enough water to float steamers in the shallows. The lucky Klondike miners who had started the stampede in the summer of 1897 had come outside by way of the Yukon River. It then became the way to go inside for stampeders who were wealthy enough to pay for the long passage. It also was used by well-to-do tourists who wanted to see what the excitement in Dawson was all about. And so world travelers Mary E. Hitchcock and Edith Van Buren went to the Klondike via the Yukon River route for their 1898 touring season.

Touring on the Yukon River Route

Mary Hitchcock and Edith Van Buren left San Francisco for the Klondike on June 15. They brought with them ample financial resources as well as

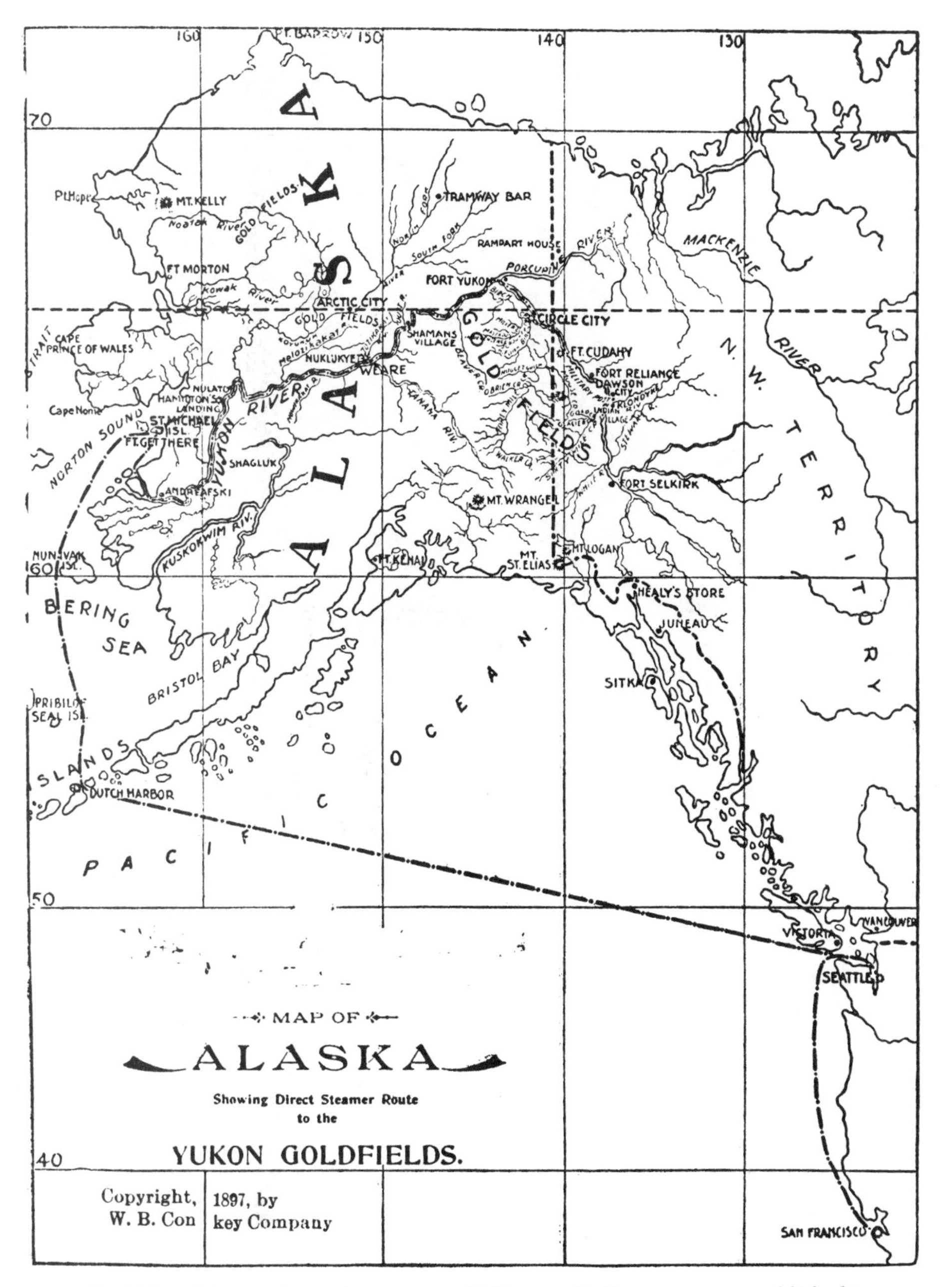

3–1. The Yukon River route as shown on an 1897 map. Similar maps were published in newspapers at the time. (U.W., LaRoche, 1897, 9075)

3–2. Edith Van Buren, at left, and Mary E. Hitchcock on board the *St. Paul*, June 1898, with their great danes Queen and Ivan. (U.W., Dow photog., Hitchcock, 1899, 9069)

an impressive array of paraphernalia. Their entourage included two great danes, a parrot, two canaries, two dozen pigeons, a large circuslike tent, and a huge store of culinary delicacies. They also had a mandolin, zither, ice cream freezer, bowling alley, and an early film projector called an animatoscope or magic lantern. It is not clear whether all this was intended simply for their own amusement or for resale at a profit in Dawson. Hitchcock's account of their journey, *Two Women in the Klondike* (1899), is both entertaining and a good description of the all-water route, albeit from a somewhat unusual perspective.

Their ship steamed across the Pacific to Unalaska on the Aleutian Islands of southwestern Alaska, then into the Bering Sea and to the mouth of the Yukon River in Norton Sound on the central west coast of the territory. There, at a small island settlement, St. Michaels, they transferred to a shallower draft river boat to continue the additional thirteen hundred miles upstream to Dawson.

By Mary's own account, she and Edith received and expected preferred treatment on the voyage. They enjoyed beds and private rooms on their ship, the *St. Paul*, when many other passengers had neither, and felt offended if men bowed but forgot to tip their hats. Hitchcock was very crit-

3–3. Saint Michaels beach as it looked in June of 1898. Here the riverboats began their journeys up the Yukon River. (U.W., Forbes photog., p. 2, 9073)

ical of customs regulations which inconvenienced her, and thought customs officials discourteous, an affront to her genteel Victorian sensibilities. She wrote, "if ever women do have their rights, and should I have a little brief authority, my first movement would be to have 'packers' on the wharf to soothe the injured feeling, smooth out the wrinkles [of searched gowns], and repair damages done by this insulting search."[1] Her expectation that others should accommodate themselves to her are also clearly illustrated by her comments upon arriving at St. Michaels.

> Those who had visited the native quarters advised us by all means to avoid them. Old Alaskan travellers on board told of a dance that could be seen, by crawling through a hole and then dropping into a cave. The dancers enter from a subterranean passage, and perform until exhausted. The greatest objection to being one of the audience (we were informed) is, that one reaches daylight with clothing so covered with vermin that it is unfit for further service. I innocently asked if we might not hire the dancers to entertain us in open air, but was laughingly told that underground performances would not be appropriate to such a changed surroundings.[2]

With such an attitude, it is not too surprising that others' accounts of her visit often contain more than a hint of resentment.[3]

Nevertheless, Hitchcock's record of her journey is full of valuable information and sensitive observations. She shows genuine concern for the welfare of those whom she found ill or grieving for lost companions or family in Dawson. And her descriptions of the physical environment of the Yukon River route are often compelling. Save for professional writers, she is rare in this respect among those recording their impression on the stampede trails. For example, the following was written upon entering the iceberg-laden waters of the Bering Sea, June 24, 1898.

> After luncheon there was great excitement, and the upper, or shade deck, was crowded by many who gazed upon icebergs for the first time. Down they floated towards us, singly, and in fantastic shapes. We steamed through them carefully—then the pulse-beats of the engine were slowed, as we saw in the distance what seemed to be an impenetrable barricade, and we began to realise the meaning of the old saying, "We could hear ourselves think." A man was sent aloft to indicate a passageway. To our inexperienced eyes, that long wall of ice before us seemed to shut out all hope of entrance, but the sailor guided us to a narrow doorway through which we passed into a clear sea. Not for long, however, did we steam at full speed. Far in the distance a small cake of ice appeared, then another, and still another, until we were soon in what could only be termed an ice-field, with the stillness of death around and not even the voice of a bird calling to its mate to be heard. It is difficult to describe the solemn stillness which pervades this vast region, dotted with ice-floes speeding noiselessly to destruction; the silence unbroken by a single sound save the throbbing of the steamer as it advanced slowly through this wilderness of space. Language becomes too poverty stricken to express the awe and admiration which fill the soul at such a time.[4]

Hitchcock's insulation from the real day-to-day worries of survival gave her the time and perspective to be sensitive to the beauty of the North Country and to record it. On other routes, stampeders of more modest means barely had the opportunity to jot more than a few notes before collapsing into exhausted sleep at the end of each day.

Once at St. Michaels, the *St. Paul*'s passengers waited thirteen days for a riverboat, for although the Yukon River's ice had gone out, the water was unusually low and many vessels were stuck on sandbars.

Such delays on the Yukon were not unusual. In fact, some eighteen hundred unlucky stampeders who had started for Dawson late in the previous summer had found themselves and all their goods stranded by the

3–4. *Alice* in the ice. Stampeders who sailed from San Francisco in September 1897, were among those stranded below Fort Yukon when the river froze. (A.M.H.A., A. A. Martin photog., B74.1.2)

ice at various points along the river. Of these late summer stampeders, only forty-three reached Dawson in 1897, and thirty-five of them had to return immediately to the outside since they had no outfits to take them through the winter.[5] It wasn't until June of 1898, by the light of the midnight sun, that the Yukon River began delivering its waves of stampeders again, many of them women and children.

By July 7 Mary and Edith were underway again, taking up accommodations on a newly built barge being towed by the river steamer *Leah*. A typical day on the river began with breakfast at 5:30 A.M., the sun having been bright for several hours already. The steady plying of the current was interrupted by stops at riverside settlements, by loading cord wood to feed the boilers, and by meeting other boats. Those going downstream carried Dawsonites who warned of typhoid and malaria and advised all to turn back. Those going upstream, like the *Leah*, were immediate cause for rivalry since most passengers were not tourists but real gold rushers, and all wanted to be first to the Klondike to stake the best claims. (Of course, the

3–5. A Yukon steamer pushing a barge up river in July 1898. (U.W., Forbes photog., p. 9, 9074)

optimistic racers did not know that most paying claims were already staked by July 1898.)

Passengers amused themselves on the decks or in the mess hall with reading, playing cards, talking, sewing, watching scenery, or "hunting" animals that innocently showed themselves along the shore. On hot days the ice cream freezer might even be brought out. When the *Leah* stopped to take on wood, her passengers went ashore for small excursions to pick flowers or berries, to fight mosquitoes, and to practice gold panning techniques.

Dinner, at 5 P.M., was often followed by dancing or a concert, courtesy of passengers and crew. The roof of the barge became an evening promenade deck or the dance floor for an energetic jig or polka, to the consternation of those who had decided to retire early in their cabins just underneath. Mary Hitchcock describes very few real hardships on the summer Yukon River journey except the ravenous mosquitoes, the leaking bilge which ruined many goods, and occasional hot cinders from the steamer's stack which might set a fire. The scenery along the river was sometimes spectacular, though Mary also notes that carelessly extinguished campfires (probably steamer cinders as well) set forest fires which blackened the shores.

> If the scenery yesterday was grand, what can be said of that through which we have been passing today? Mountains, and relays of mountains, narrow gorges, rapids, all that is most wild and

3–6. Mary Hitchcock, at left, practices panning gold at a stop downriver from Dawson, while other passengers await their turns. (U.W., Hitchcock, 1899, 9070)

picturesque! We had been too rapt in admiration even to read, but, as there must be ever a slight blot on all that is beautiful, so this scene was partially marred by the gradual approach of a heavy fog, as we thought it, until the air became laden with smoke, and, as night came on, we saw that the mountains on all sides were on fire. Truly a gorgeous sight, which would have been still more brilliant had it not been for the ball of fire that hung in the west, making all else insignificant by comparison.[6]

When at last they arrived on July 27, 1898, after twenty days of river travel, Van Buren and Hitchcock set up their huge tent across the river from Dawson and began a round of sight-seeing and entertaining. Two months later, having disposed of much of their fantastic baggage, they left September 23 for Whitehorse on the steamer *Flora* with an anxious eye on

3–7. Steamer *Seattle No. 1* arrives at Dawson City in June 1898. This photo was taken with the light of the midnight sun. Arrival of the first river steamers of the summer, loaded with long-awaited supplies and letters, was a cause for celebration in Dawson. In the upper center of the photo is the white scar of a prehistoric landslide, referred to as "Moosehide." (U.W., Hegg photog., 740a)

the falling snow and the prospect of being frozen in before completing their upstream journey to Lake Bennett and the outside.

Lake Bennett formed some of the headwaters of the Yukon and lay only thirty miles from the coast of the Inside Passage. Mary and Edith got through, however, and hiked out over White Pass. Just across the pass, in White Pass City, they became passengers on the newly built narrow-gauge railway. In response to the pressures of the stampede, the railroad from Skagway, at the coast, to Lake Bennett had been under construction at a furious pace. Mary's description of some of the old miners' reactions to the train conveys clearly the significance of this railway to those who had fought the hazardous White Pass and Chilkoot trails, which the train now replaced. These were probably not stampeders but prospectors who had been exploring the north country for years and now, having finally struck gold in the Klondike, were returning to civilization.

At last the whistle sounded. "All aboard!" was shouted. Then the Klondike "boys" began to exclaim joyously, "A train at last after all these years!" "How long since you been in one, Jim?" "Too long to talk about," said Jim, as the tears rolled down his weather-beaten cheeks. The "boys" began to sing *Home Sweet Home*. "My old mother don't know I'm a-coming. Poor Bill! his people have all died, he's been away so long, and he ain't got even a sweet-heart to welcome him back, but he'll have a hot time in Skaguay to-night with all his nuggets disappearing." The "boys" caught up the strain and *A Hot Time in Skaguay* was predicted from dozens of throats.

Some stiff, stately persons seated in front of us drew themselves together, their noses high in the air, and gazed contemptuously upon the noisy rabble. They could not see the pathetic side of the picture—of how the poor "boys" had tramped, footsore and weary, for days, months and even years; putting up with privations of all descriptions, suffering from lack of proper nourishment, half frozen in winter or risking their lives in going to the assistance of a less fortunate comrade, or they would not have frowned upon those shouts of joy at being once more within the bounds of civilisation. A sudden whistle! "A cow on the track, boys! let's get off an' look at her. I've forgotten how one looks."[7]

Working the Yukon River Route

In contrast with the relatively well-to-do tourists, Nettie Hoven arrived in Dawson by river on July 26, 1898, with little money but a lot of pluck. Nettie was the sort of person—male or female—who won admiration in the gold rush. The account of her seven-month journey that commanded the pages of the *Klondike Nugget* is thrilling, a type of adventure literature that filled the journals of the day.

"It is very doubtful if any other woman ever had a more varied experience that Miss Hoven has been through since leaving New York on the 16th day of December last," the *Nugget* story began. She started out as a worker on the steamship *Columbia* bound for Seattle, "which for months before had been widely advertised to bring Mrs. Hannah Gould's party of 150 widows to the Klondike.[8] Miss Hoven, however, was not one of the widows. She was coming on a little expedition of her own. . . ." But the account that followed was hardly of a "little expedition."

"Our trip" said Miss Hoven in relating her experiences, "was a very pleasant one until we reached the city of Rio [de] Janeiro.

Here we were compelled to lay over for ten days to repair some broken parts of the ship's machinery. After leaving that city we proceeded down the coast and were wrecked in Smite's Channel which runs through the Straits of Magellan. For three days and nights we remained upon a rocky island in the vicinity of a cannibal village. The natives were nearly seven feet high and very savage and we were constantly in fear of an attack, but fortunately escaped. The officers succeeded in repairing the ship sufficiently to enable us to leave although she was leaking in seven different places at the time of our departure. The holes were stuffed up with rocks and hay and thus we managed to keep afloat.

"After rounding the Horn we headed for Valparaiso. Here we ascertained the fact that the company controlling the 'Columbia' had failed and the captain was without funds to secure necessary provisions. The passengers then took matters in their own hands and raised a sum amounting to $28,000 for the purpose of repairing the steamer and procuring provisions. We remained in Valparaiso forty-seven days and then left for Seattle which city we reached on the 29th of April. On May 2nd passengers and crew alike were sent ashore without provisions or money and told to shift for themselves. Nearly all had paid in advance for outfits which were to be ready for them at Seattle together with a steamer to convey them to Dawson. Outfits and steamer alike failed to materialize, and as a result nearly all my fellow passengers on the 'Columbia' are still stranded in Seattle."

"How did you manage to get passage to Dawson from Seattle?" queried the *Nugget* man. Miss Hoven laughed. "Well, it took a whole lot of nerve," said she, "but I fixed it all right and I'll tell you how I did so. I was walking along the water front and saw the sailing vessel 'Hayden Brown' just ready to pull out for Kotsebue Sound. I knew she would stop at St. Michaels and I would take chances on getting from there to Dawson. I secured my grip and just as the ropes were being pulled in, walked aboard. No questions were asked until we were well out and then there was nothing for them do but allow me to come. They treated me very well and gave me some work to do to help pay my passage. The ship was poorly provided with food, however, and nearly everything was exhausted before we reached Dutch Harbor. There were 187 passengers aboard and less than $700 worth of provisions. At Dutch Harbor the passengers raised $1000 and with this sum purchased supplies which enabled us to reach St. Michaels.

"On board the Hayden Brown were forty-three men and one woman who had been in charge of a man named Chase. Chase had contracted to provide them outfits and land them at Dawson for sums ranging from $60 to $110 each. The party was sent ahead on the 'Hayden Brown' and Chase himself agreed to follow on another ship with the outfits. Neither Chase nor the outfits ever arrived at St. Michaels and most of the party are still there without money or supplies. A few came up in the 'Sovereign' to Dawson.

"When I was ready to leave St. Michaels, I went to the captains of several boats and asked to be allowed to work my passage to Dawson. They all refused until finally I found Captain Danaher of the 'Sovereign.' After hearing my story he told me to bring my things on board. He as well as the other officers treated me very kindly and the same privileges were given me on the whole trip as their passengers enjoyed.

"Altogether I have no reason to complain of my trip. I started out to reach Dawson and have traveled 13,000 miles to do so, but here I am. I left New York without a cent and reach Dawson with money in my pocket so I think I ought to be satisfied."[9]

It makes an interesting story—plucky, attractive, young woman uses wits and courage (and feminine charm?) to get to Dawson. But Nettie Hoven didn't tell one very important part of her story, and understandably so. She was running from a man, John F. Mellen, with whom she had lived in New York. Mellen had, in fact, been on the boat with her around the Horn and threatened to kill her. She apparently had given him the slip in Seattle. But even as Nettie was telling her story to the *Nugget* reporter, Mellen was finding a way to follow her to Dawson.[10]

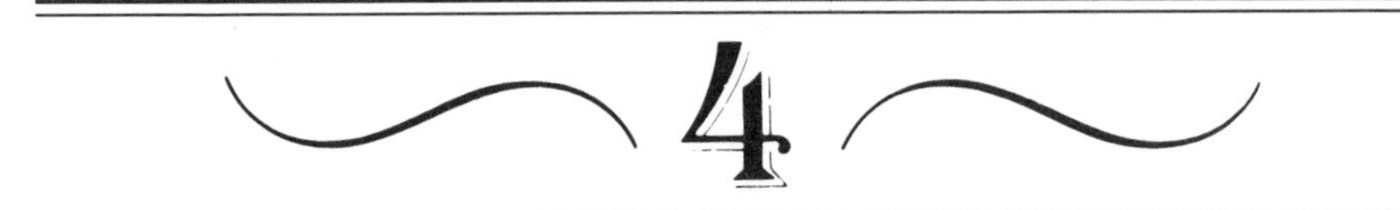

TREKKING THE CHILKOOT TRAIL

"For God's sake, Polly, buck up and be a man"

THE DYEA and Skagway trails were the beginning branches of what was, by far, the most popular route of the Klondike stampede, the Overland trail.[1] Not only did the Overland trail carry the most stampeders but it is also the route for which we have the most written and photographic records.

The heads of the Dyea and Skagway trails were only six miles apart on the glacier-rimmed north end of the Inside Passage, Lynn Canal. Each required trekking overland along a river valley to cross a pass through coastal mountains. The Dyea trail, which still exists as part of the Klondike Gold Rush Park, is thirty miles long. The defunct Skagway trail is now paralleled for much of its length by a portion of the Klondike Highway, which runs between Skagway and Whitehorse. The Skagway trail covered forty-five miles before joining the Dyea trail at Lake Bennett or Lake Lindeman. These lakes are some of the headwaters of the Yukon River, which flowed past the new townsite of Dawson, 560 miles to the north.

Though one could continue overland in the winter from Bennett to Dawson, most stampeders chose to wait until summer and use the river for transportation. As a result, the so-called Overland trail was actually only 5 percent by land. Stampeders built boats, loaded them with all their belongings, and headed downstream through rapids, sandbars, lakes, and some of the most beautiful wilderness scenery in the North.

Dyea to the Canyon

The Dyea or Chilkoot trail had been used by generations of coastal Native Americans, the Tlingits, who controlled trade into the interior. When

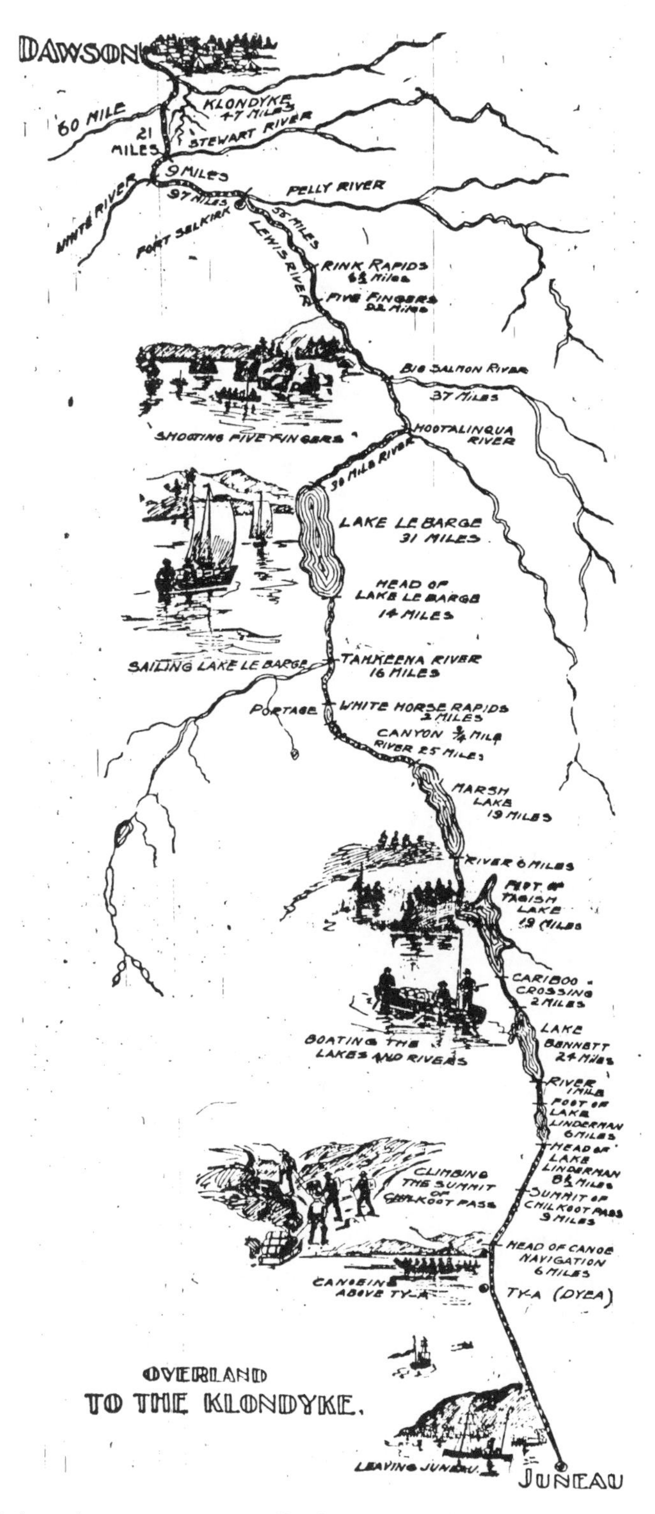

4–1. Sketch from the San Francisco *Call*, July 23, 1897, illustrates the Overland trail. Here the Dyea/Chilkoot branch is used.

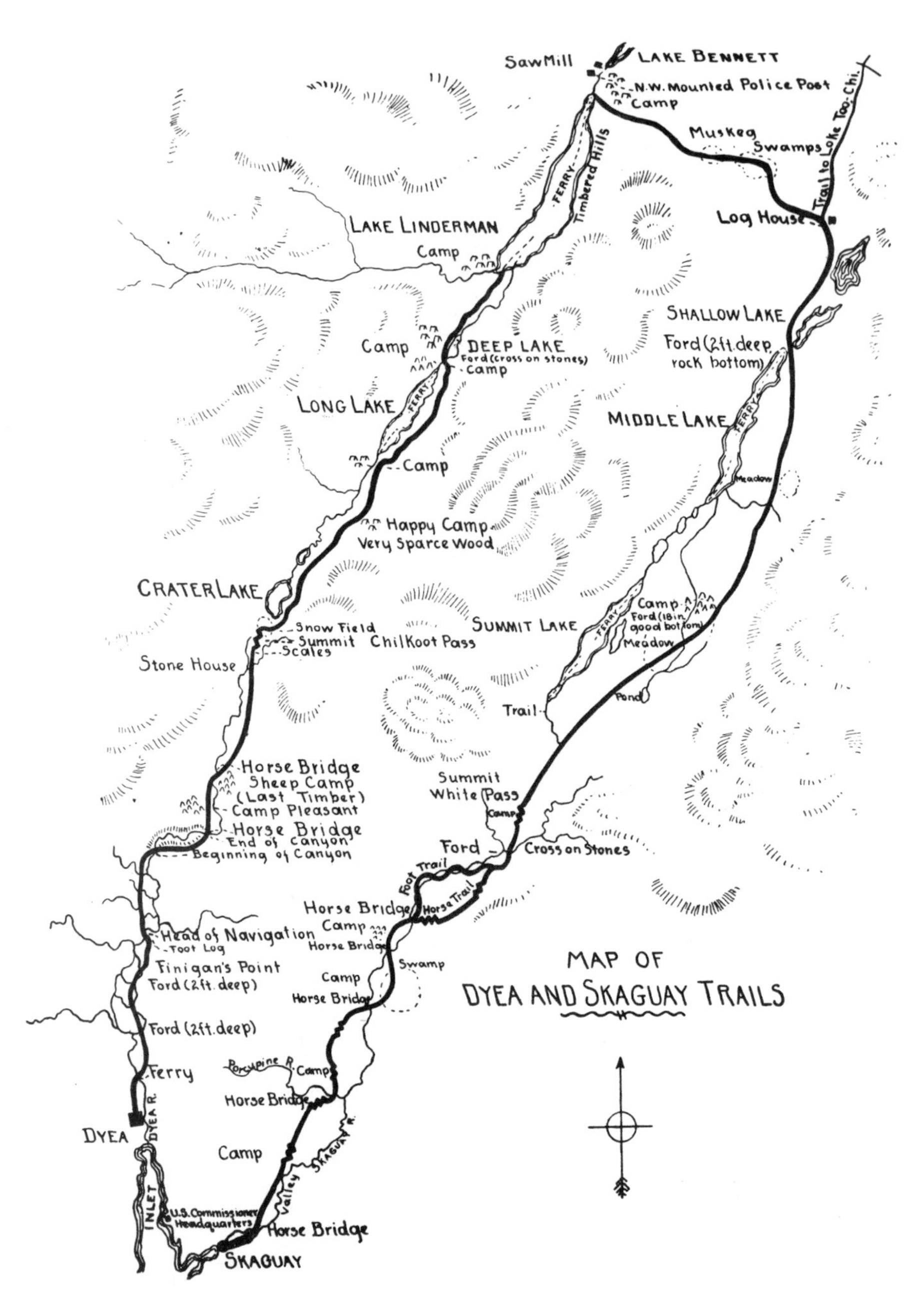

4–2. Map from gold rush times shows both the Dyea/Chilkoot and Skagway/White Pass trails. (U.W., LaRoche, 1898, p. 22 + 1, 9076)

4–3. Actress Esther Lyons, well-known on the New York stage in the late 1880s and early 1890s, became Cad Wilson for her career on the Pacific Coast in the mid to late 1890s. She was one of Dawson's most popular entertainers. (A.H.L., Wickersham coll'n, PCA 277–1–195)

Caucasians from North America and western Europe arrived in increasing numbers in the nineteenth century to explore, trade, hunt, and mine, they followed the well-known Indian trail. One such early party was that of Veazie Wilson. In 1894 he contracted with *Century Magazine* to write a series of articles about the Yukon River Basin, whose mineral wealth was just beginning to be discovered.[2] So he set out on a photographic expedition across Chilkoot trail and down the Yukon. He was accompanied by his wife, Josephine, and by an actress from the New York stage, Esther Lyons Robinson, and her husband.[3] As secretary for the trip, Esther Lyons later helped Veazie prepare a Yukon River guidebook,[4] and she wrote and lectured about the Yukon herself, as well.[5] Four years after their expedition, without revealing her identity, Esther Lyons returned to the Yukon as "Cad Wilson." She was celebrated as one of Dawson's most popular stage entertainers in the height of the Klondike boom.[6]

Other early travelers over the Dyea trail who would later figure prominently in the Klondike were George and Kate Mason Carmack, the original locators of the Klondike gold. Before the discovery they both packed supplies for others from Dyea in order to make some money. In the winter Kate earned a little more by sewing moccasins and selling them to prospectors.[7]

4–4. The coastal side of Chilkoot Pass on a clear day in 1898. Stampeders with their gear gather at the foot of the pass to prepare for the final ascent. The dark, solid line to the left of center is formed by human packers, one after the other, climbing icy steps up over the pass. The light, sweeping lines just right of the main column are the channels made by packers sliding to the bottom to pick up another load. To the far right are those lured by the gentler appearance of the Petterson trail, which is longer and more difficult on the Canadian side than the traditional Chilkoot Pass trail. (U.W., Larson coll'n, Hegg photog., 105c)

Even as late as 1897 when the Klondike stampede began, the Dyea was one of only two well-established trails into the interior, the other being the Yukon River route. But because the Dyea trail was quicker going into Dawson and could be used year-round, it was the one preferred by most early stampeders. (Soon after the rush began, new alternatives, such as the Edmonton or Skagway trails, attracted some of the thousands heading for Dawson by whatever way they thought they could get there.)

Although the Dyea trail was shorter than the new, neighboring Skagway (White Pass) trail, it did have one feature that was intimidating—the precipitous Chilkoot Pass over which pack animals could not be taken. Many a stampeder opted for the Skagway trail solely to avoid Chilkoot Pass. And for those who did take the Dyea trail, crossing Chilkoot Pass was often their most enduring memory of the gold rush, whether he or she eventually made it to Dawson or not.

Early rushers over Dyea trail had no wharf at which to land. They began their journey by being lightered to Dyea beach, where they were dumped with all their outfits. Nearby were a Native American village and the Healy and Wilson Trading Post, founded in 1885. Native American women and children were thus at Dyea long before the gold rush of 1897–98, as was Mrs. Healy, who helped run the trading post.[8] Once the rush was on, a town quickly grew up around the store.

By the summer of 1898, Dyea had false-fronted wooden buildings, wharves, and telephones. Excited stampeders slogged through muddy streets, patronizing the usual assortment of stores and businesses established to supply their needs—food, clothing, hardware, and animal feed stores; restaurants; bars; hotels; bakeries; packers for hire; jewelers; and prostitutes.

Within one more year Dyea was nearly deserted. A railroad had been built from nearby Skagway over White Pass. As a result, going into or out of the interior was a lot easier, and few thought it worthwhile to struggle over Chilkoot Pass.

Moving an outfit up the Taiya (Dyea) River Valley took persistence. Although dehydrated food and careful planning reduced the weight of supplies considerably, each person still had to bring in 1,000 to 2,000 pounds of goods to comply with the Canadian government's requirement for one year's provisions. Every means was used to convey these tons of supplies up the trail. Horses, mules, donkeys, ponies, professional packers, dogs, and even cattle could be hired for transportation. In the winter, individual stampeders or animals could pull sleds, and in the summer wagons or small handcarts could be used on flatter portions of the trail.

Many stampeders carried every pound on their backs, however, because they couldn't afford anything else. Despite loads as heavy as seventy pounds, a person traveling alone had to make fifteen to twenty-five round trips to carry in 1000 to 1500 pounds of provisions. At this rate, even the

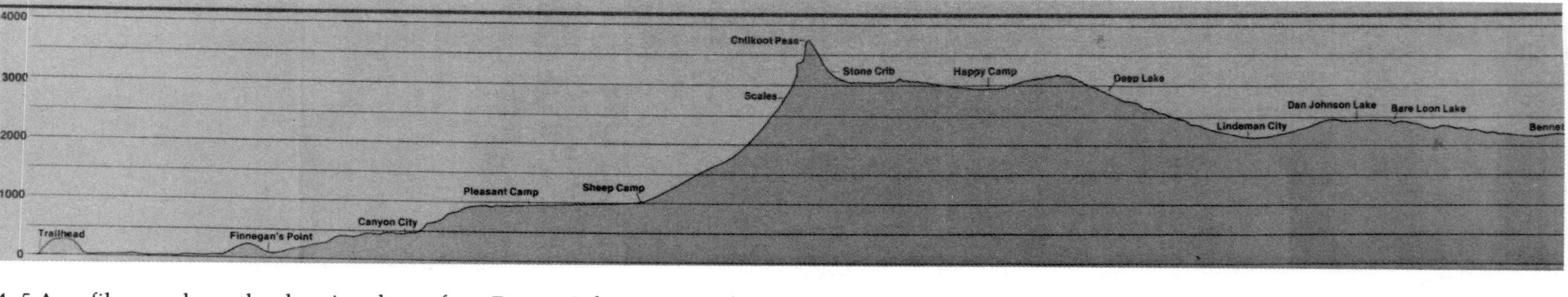

4–5 A profile map shows the elevation change from Dyea to Lake Bennett and gives an immediate impression of why Chilkoot Pass, at center, was so intimidating. Vertical scale is in feet, approximately. (U.S. Park Service, 1986)

4–6 The town of Dyea looking northeast. The Taiya (Dyea) River's two branches are seen in the foreground and background. Beginnings of the beach are to the right. (U.W., Larson coll'n, Hegg photog., 51b)

4–7 Stampeders, including several women in foreground, sort through their goods on the beach at Dyea, the head of the Dyea/Chilkoot trail. The photo was probably taken in late 1897 or early 1898. (U.W., Winter and Pond photog., 325)

4–8. Trail Street, Dyea, probably in early 1898. Muddy streets were very much a part of town life in both Dyea and Skagway. (U.W., Hegg photog., 52)

flattest first six miles of trail could easily take five days and two-hundred-forty miles of packing.

Here the advantage of teamwork, whether between family members or friends, became obvious. Stampeders without partners were more likely to become ill, because they didn't have the time to look after themselves. They might forego good meals and warm, dry clothes at the end of the day, because they were too tired to do anything but mix flour and water for pancakes and fall into an exhausted sleep. When people traveled with others and worked together, however, sleds could be more heavily loaded, equipment could be shared so fewer pounds were needed per person, and camp duties could be done by one while the others continued packing as long as possible.

Like Emily Craig on the Edmonton water route, Inga Sjolseth (Kolloen) kept a diary as she went to the Yukon on the Dyea trail in the spring of 1898. She was a single woman at the time, a recent immigrant from Norway, and she traveled with a group of friends from Seattle. Inga re-

4–9. A family with an infant sets out from Dyea in April of 1897, assisted by a young Indian packer. The woman is most likely Ella Card carrying her infant daughter. Fred Card, Ella's husband, may also be among the men shown. Their cart, piled high with furnishing, was probably handmade at Dyea. (A.H.L., Winter and Pond photog., PCA 21–13)

corded many examples of the benefits of their successful partnership in her diary. And when her friends alone were not enough, Inga's Christian faith was a source of strength and comfort to her.

Mon., March 21 We got up at five this morning, had a little coffee first and then S. [Sandvig] and A. [Andersen] went for a load of our supplies. When they got back, we prepared breakfast and the men went after another load. I went along and pulled three sacks on my sled, and it went pretty well. There are hundreds of people here dragging or carrying their supplies, all striving to reach the Klondike. Some have horses and still others have dog teams, but most of them act as their own horses. I have baked bread and biscuits and cake, and today yellow peas were the only thing I had. Our stove does not bake well, so I had to keep turning everything in order to get it baked. We had chicken for dinner, which is a luxury here in Alaska.

Tues., March 22 We were up early today and dragged a load of our supplies to the next stopping place. I dragged one sack this time because I was so weak that I could not do more than that. When

4–10. Stampeders drag some of their goods on sleds across mud and snow of Dyea trail, 1898. (Kermit Edmonds coll'n, B. W. Kilburn photog.)

4–11. Inga Sjolseth (Kolloen) in about 1898. (U.W.)

we got home I was tired, and after we had eaten I became quite uncomfortable, although I remained on my feet until six o'clock. Then I went to bed, but did not go to sleep until nine o'clock. They played and sang for us which I thought was delightful and uplifting, especially the songs which were sung to Jesus Christ and which turned our thoughts toward Heaven. My conviction and my constant thought is that in God I am happy and contented. There has been a strong wind today and some snow has fallen.[9]

Many of the coastal Native Americans earned good wages packing outfits on the Dyea trail and over Chilkoot Pass. This had been the practice ever since Caucasians had first used the trail. (See Ogilvie's account of his 1887 expedition, for example.[10]) But during the gold rush, whole families went to work, with many a stampeder surprised by the seventy-pound loads that the women and children could carry on the rough trail. At the same time, many complained of the high cost of packing—forty cents per pound from Dyea to Lake Lindeman in September 1897. The $400 fee for each one-thousand pounds was in 1897 dollars, of course, when a whole outfit would cost about $300 to buy.[11]

Though familiar with the Dyea trail, the Native Americans, like the stampeders, were also subject to its perils—sudden blinding snowstorms; cold, driving rains followed by floods scouring the valley floor; and avalanches from snow-laden, steep mountainsides. One Native American woman, separated from her party on the trail in a snowstorm, used all her meager resources to save the life of her baby. A search party was sent out from Dyea when they were discovered missing. Someone noticed a slight but unfamiliar mound of snow near the trail, so the searchers dug down and found the woman. She was dead, but she had scraped a small hole in the snow, placed her baby wrapped in her own parka in it, then huddled over the infant, protecting it with her body. The baby lived and was taken to Dyea, where it was adopted by another family.[12]

Belinda Mulrooney (Carbonneau), an enterprising single woman in her late twenties, was among the first to cross the Chilkoot Pass in the early spring of 1897, even before the first gold-filled boats had reached West Coast ports and triggered the stampede. She had heard of the discovery while running a mail-order business in Juneau in the summer of 1896 and while working on a steamer plying the Inside Passage. She recognized her chance to make good and capitalized on the opportunity. Once in Dawson, she became one of the wealthiest in the Klondike, holding a number of claims and managing several businesses. She also became very trail-wise and recalled, with a touch of humor many years later, how even in relatively good weather an unprepared traveler is threatened by cold and exhaustion.

> If you start to freeze on the trail you have the fight of your life on your hands. You want to lie down, for the snow looks like a feather bed.
>
> I remember once on the trail there was a nice-looking woman from Seattle lying in the snow. Her husband had gone ahead to Dawson and said for her to wait in Skagway, but she was impatient to be with him. She started in on her own, poorly dressed and with her one little sled. When I came on her I said: "Come on. You'll have to get up and make an effort." She replied: "Leave me alone. Leave me alone. It feels wonderful." I took the pull bar

4–12. Belinda Mulrooney (Carbonneau) packs her outfit of commercial goods through Dyea Canyon in April of 1897. She later owned and managed a number of businesses and mines in the Klondike and became one of its wealthiest women. Others in her party, from front to back, include Williams, John Lee, Dan Fraser, Bob Menzie, Gus Biegler, Ed Hutchison, Williams, Clare M. Gillett, Marlin Mosier, and Bert Bower. (A.H.L., Winter and Pond photog., PCA 87–682)

out of my sled, put it under her and rolled her on to my sled. I had to tie her on good and hard to keep her from climbing off. Then I gave the sled a good kick down the hill to my camp.

Bill McPhee and another old-timer were huddled in my tent. "What are you doing in here?" I said. "I've got a poor woman here and I've got to thaw her out."

McPhee (a well-known Dawson innkeeper) said, "Belinda, put her in here between us. She'll be as safe as if she was in God's pocket!" So I did, and when we got down safe to Dawson and she found her husband, Ole, a Norwegian, she said "Honey, Honey, come and take a look. Those are the men I slept with on the trail!"[13]

The Canyon to Sheep Camp

Six miles from Dyea, the Taiya River runs through a canyon. At this point the trail in summer leaves the river and becomes steeper and rockier, passable to horses but not to wagons. In the winter, the frozen river was the trail, but what had been log-entangled waterfalls became cliffs of ice over which goods had to be carried by hand.

Despite the rough passage, this section of the trail is very picturesque, with tall trees and occasional views of glaciers hanging in high valleys to the west. But during the stampede years, the trees perished, supplying shelters or sleds and fueling campfires. With time, the beaten path began to look truly beaten.

Past the canyon the summer trail again returns to the river's edge, arriving eventually at Sheep Camp, about four miles from the summit. This is the last camping spot below timberline until several miles beyond the pass, so during the gold rush Sheep Camp quickly grew into a little city where stampeders gathered their wits and energy for the long climb over the pass. Inga Sjolseth (Kolloen) described early spring travel on the trail above Canyon City and conditions at Sheep Camp.

Tues., March 29 Today we moved from Canyon City to Sheep Camp. The trail in from Canyon is narrow and crowded, lying between two steep mountains. If one looks up one sees only mountains over one's head. The snow has turned to water, so progress has been very difficult. Ida [Bodin] and I pulled a loaded sled together through the canyon and heard many comments about it. We heard one man say he wished he had had such a "tram" as we were. Some of the men took off their packs and laughed at us. We have now set up our tent here in Sheep Camp.

4–13. Stampeders camp at the foot of Dyea Canyon in the fall of 1897 with a lean-to tent improvised from paisley cloth. (U.W., LaRoche photog., 2026)

4–14. Two children watch as men push a sled up a frozen waterfall, Jacob's Ladder, in Dyea Canyon, April of 1897. This may be the Ed Kelly family. (A.H.L., Winter and Pond photog., PCA 87–675)

4–15. Pack train returns for a load down the boulder-strewn Dyea Canyon in the fall of 1897. Trees are already showing effects of being exploited for lumber and firewood by stampeders. The destruction shown here may also be the result of a flood that swept down Dyea Valley in September of 1897. (U.W., LaRoche photog., A)

4–16. Packers on trail near sheep camp in fall of 1897. (U.W., LaRoche photog., B)

> The snow is thick under our tent. A little rice with berries on it. Then we are ready to sleep.
>
> *Wed., March 30* It has snowed and rained all day and the water has come through our tent so many of our provisions are wet. It is uncomfortable and unhealthy to live in a tent under these conditions. We must hope that real soon this will be better. The snow has melted, and I fear that the tent may fall down any minute. It leans to one side. I have visited Miss Knudsen today, also Mr. and Mrs. Drange. I also visited Mr. Waldal, Mr. Knudsen and Mr. Melseth. Mr. Paulsen was here with us.[14]

Throughout her diary on the Dyea trail, Inga mentions people she has met, acquaintances she visits, helping others with their work and receiving help in turn, and social calls at the end of the day. For her, trail life had a very large social component which may reflect not only her own temperament but also the cooperative, communal orientation of her Scandinavian friends. In addition, her role as camp attendant rather than packer probably gave her more opportunities and reasons to meet some of the large numbers of people who were on the trail that spring.

Also at Sheep Camp in the spring of 1898 were C. J. and Ethel Berry, the lucky Fresno couple who had struck it rich on Eldorado Creek. They were coming back now with new supplies for another summer season of work at the claim. With them were several men and Ethel's teenage sister, Alice Edna Bush, or "Tot" to her family. Tot was thrilled to be along on the great adventure and fascinated with everything she saw. Later in her life she wrote about Sheep Camp.

> I was delighted when we pitched camp on the main street, for I had been afraid C. J. would pick a place 'way back and I should miss seeing things. Everything was new to me and most exciting. We had one tent for cooking and one for sleeping. Mind this—all of our provisions and equipment, tin cups, tin plates, spoons, a sheet-iron stove and a handful of cooking pans, would have to be pulled miles and miles over frozen trails. . . .
>
> I liked Main Street as we were almost opposite a dance hall run by two women, and I'll say that they could take care of themselves. They could dance, sing, swear, play roulette, shake dice and play poker; in fact, they could do almost anything, and as I had never seen anybody like that before, I kept my eyes on their dance hall every minute. Of course, C. J. wouldn't let me go inside the place.[15]

But it was not all fun and games for Tot. She and Ethel were in charge of camp duties, so there was plenty of work to do. And though Tot liked all the activity at Sheep Camp, she also had to put up with very little privacy in the crowded camp.

> It was hard work cooking for so many with the same kind of food every day and so few utensils. Ethel and I did the cooking for our crowd. We had to set up everything twice for each meal, as we didn't have enough dishes. We had brought one sack of potatoes; when these were used, we should not eat potatoes again until we returned to the States. The cooking smelt so good to strangers passing the tent that they would poke their heads in to ask if it was a boarding-house.
>
> The tent where we dressed had a double flap on the entrance, pinned with six-inch safety pins. Even with that one of us had to stand guard while the other bathed and put on clean clothes. Someone would work at the door trying to get his head in, just to see who lived there. We would call out: "You can't come in! This is private." Then he would ask: "What the hell is that?" One day when a man was talking back to us at the door and annoying us, Clarence stepped up and landed a good blow on the fellow's chin, saying: "Now I guess you know you are not wanted here."[16]

Still, not all the men were so rude, and Tot was interested in them. Yet she found that they seemed to show very little interest in her.

> I often wondered why none of the many men at Sheep Camp seemed to take any notice of us; plenty of them were young, and there was only a handful of women. Perhaps the fact that we had eight men explained their indifference, or perhaps it was that certain look Clarence could give; or perhaps a man can have only one kind of fever at a time, and they surely all had the gold fever. Everyone's thoughts were centered on one thing, to go as far as we could each day and to know that we were one day nearer our goal.[17]

Mae McKamish Meadows, the Santa Cruz woman who had written so enthusiastically to relatives from Juneau, arrived at Sheep Camp in September of '97, several months before Inga Sjolseth and the Berry party. Her perspective of the settlement was different still, because of both the timing as well as Mae's temperament. Mae and her husband Charley Meadows were a couple who painted life with broad strokes. "Arizona Charley" had been a star performer in Buffalo Bill's Wild West Show and was known as a frontiersman, Indian fighter, scout, and crack shot with the rifle. In the 1890s Charley had started his own show in which Mae was one of the featured horseback riders and chariot racers. When they left for the Yukon, Charley declared that they were "outfitted perfectly and prepared for anything we might meet."[18] Among other things, they transported a general store, a restaurant, and a bar up the trail, which they set up at each stopping place to sell drinks for ever-increasing prices.

4–17. Mae McKamish Meadows in about 1895. (Jean King coll'n)

Like Inga Sjolseth, Tot Bush, and Ethel Berry, Mae stayed at Sheep Camp while her party's outfit was being packed over Chilkoot Pass. But Mae was there in the fall of 1897 when the camp was flooded and wiped out by a mud slide.

> We were going to move to the foot of the summit that day [September 18], so had everything there, except my clothes, and they are perhaps to Frisco by this time. The river came up into the big tent about 2 o'clock in the night. Charlie and I were sleeping on the ground in the corner, and we woke up to find a foot of water in the tent—up to our first blanket, and the wind blowing a perfect gale. Billy and Anson were sleeping in some boxes. Charley

4–18. Arizona Charley Meadows in about 1892. (Jean King coll'n)

went to the saloon and made me a bed on the bar (the only high, dry place any where), with the only two dry blankets left. I went to bed this way. The flood was at 6:30—was caused by a land slide at a place called the Stone House, half way to the foot of the summit. The rest of our people were up there. They heard the slide, and next day they heard Charley and I were drowned. In fact, it was reported all over Sheeps Camp in a minute that we were seen to go down the river. People saw the flood coming—could see a black dust and trees falling every where. No one thought it would be so bad.

Charley went out, took a look; came and told me there was a snow slide coming, but that I had plenty of time. I got on part of

> my clothes. He came running in, and said the flood was there, and to run for my life. Well, I grabbed the rest of my clothes and a little grip. I told him to take some of the bed if he could, which he did. We just got out to the front, when the slide struck the back of our tent. We were about three yards away when the tent went down. Well I wanted to stop to see the water, and try to button my clothes, but Charley kept saying to go on farther up, so I had to keep on running. If we had not got out at that moment we would have been drowned. It was terrible. Charley said if he had a Kodac [*sic*] of me as I was running from the Sheeps Camp flood, there would not be any use of going to the Klondike, as that would be a gold mine itself.[19]

The high-spirited Mae and her flamboyant husband Charley must have been quite a pair on the stampede trail. Whatever they did, the Meadows did with style—even after near disaster, they could still joke about their experience. And they were still determined to get to Dawson to make their fortunes. So they replaced what they had lost in the flood by buying from those who gave up and abandoned the stampede.

> Our provisions were all saved, except just the little we had there to live on. They found my nice fur coat, and tan jacket, black divided skirt, all soaked through with mud. My packing case with all of our underclothing, fur hoods, chest protectors; in fact, everything we had to wear, were all gone. Charlie lost all of his hats, his lovely new pistol, and that just broke his heart. Billy lost a fine pair of rain boots he had just paid $10 for. Baudry was not here, so he was all right. He had gone on with the rest of the crowd with the provisions.
>
> Well, you could not see the people for the mud going back to God's country. People could not get out fast enough. They sold out for anything they could get. There was a German woman going back, so I bought her underclothes, but they were not enough. She let me have them for just what she paid for them in San Francisco.
>
> That night we slept in a log house on some of our boards, with the water dripping on us all night. We found the big tent, and hunted for two miles down the river to try and find my clothes, but could not. Then we took our last load and went to the foot of the summit, which is the hardest trip of all. Mud knee deep, and great rocks. You would never think a horse could get along at all. There are dead horses every where, as the owners can not get any thing to feed them since the flood. It was dark long before we got to camp. One of our horses fell and turned over on his back with all our blankets in three feet of mud.[20]

4–19. Sheep Camp, the last stop before timberline and Chilkoot Pass. This photo was taken shortly after September 18, 1897, when the flood described by Mae Meadows swept through the camp, destroying everything. Within a matter of weeks, the clapboard town would rise again. (U.W., A. Curtis photog., 46118)

Chilkoot Pass

Although Chilkoot Pass is only 3,550 feet in elevation, the final ascent of about 1,000 feet is done within a half mile and over large boulders in the summer or through snow and ice in the winter. Because this section of the trail was not passable by loaded pack horses, all goods had to be carried by people, until a tramway was built about December of 1897.

Even so, the cost of transportation, either by tramway or packers, was high. So many stampeders packed their two thousand–pound outfits themselves, climbing the pass again and again to ferry all their goods. In the summer, they struggled over the bare rocks, using walking sticks to help balance on the uncertain footing. In the winter, steps were cut into the ice, forming a great white stairway. Not surprisingly, many items in an outfit were considered more critically at Chilkoot Pass. Many nonessen-

4–20. Approach to Chilkoot Pass, seen in upper portion of photo, slightly left of middle, is more and more rocky once leaving Sheep Camp. Here packers rest about 1.5 miles from the pass in the fall of 1897. Native American men and women both hired out as packers. (U.W., LaRoche photog., p. 38, 2038)

4–21. Packers form a nearly continuous line up to Chilkoot Pass. Dugouts for resting are to the left of line, beginning about halfway up in photo. Chutes for sliding to the bottom fan out to the right. (U.W., Hegg photog., 97)

4–22. Caches and stampeders at the summit of Chilkoot Pass. (U.A.F., Blankenberg photog., 57–1041–3)

tials were tossed out to lighten the load before climbing to the top. Once at the summit of the pass, packers cached their supplies then returned to the bottom to pick up the next load. In the winter, at least the return was quick: One simply slid down a well-worn chute at the side of the trail. Inga Sjolseth (Kolloen) described the scene at the foot of the pass.

> *Wed., April 20* We got up early, had our breakfast, and packed our supplies. We were ready at 6 a.m. Then a man came to drive us to the summit. They had a sled and a little undernourished horse. It took a long time to get up there. We sat and watched those who scrambled over. They went up very slowly, but came down the chute again with great speed! Many sat down and slid most of the way. It took only a few minutes. It went so fast that I was afraid to watch. It was so late when my supply train went over that I was afraid to go over with it. I was really afraid. So I went down to Sheep Camp again.[21]

The going was slow, not only because of the rough terrain but also because most stampeders were not used to packing. In addition, the weather at Chilkoot Pass is often treacherous. Winds funnel up the Dyea valley dumping water on the slopes of the coastal mountains as their moist air rises and cools. Even when not carrying rain or snow, the blasting air makes travel across the pass uncomfortable. When combined with sweating from exertion, there is added danger of hypothermia, even in the summer.

Martha Munger Purdy (Black)'s description of her crossing of the pass in the summer of 1898 captures some of the drama and pain experienced by many of those who made it that far on the stampede. In Chicago, a family asked Martha to act as their agent in locating the estate of an uncle, who had died leaving them $1,000,000 in Yukon gold. They had not received any of the money, but if Martha could recover it, she would get half. So Martha, her husband Will Purdy, and her brother George Munger set out for the Klondike. While in Seattle, Will changed his mind and decided to go to Hawaii (then the Sandwich Islands). He suggested that Martha either accompany him or simply return to Chicago. She wrote, "Even after 10 years of married life how little Will Purdy knew me!"[22] Martha went ahead with her mission to the Klondike. For her that summer, even the approach to Chilkoot Pass was nearly overwhelming, her misery amplified by the burden of her restrictive clothing.

> As the day advanced the trail became steeper, the air warmer, and footholds without support impossible. I shed my sealskin jacket. I cursed my hot, high buckram collar, my tight heavily boned corsets, my long corduroy skirt, my full bloomers, which I

4–23. Stampeders huddle outside the Canadian customs tent at the summit of Chilkoot Pass in a snowstorm. (U.W., Hegg photog., 108)

4–24. Packers scale the boulders of Chilkoot Pass in the fall of 1897. (U.W., A. Curtis, photog., 133a)

had to hitch up with every step. We clung to stunted pines, spruce roots, jutting rocks. In some places the path was so narrow that, to move at all, we had to use our feet tandem fashion. Above, only the granite walls. Below, death leering at us.[23]

Years later Martha could recall vividly that the final steep ascent over the pass took all the strength that she could muster. Melting snow and rotted ice added to the hazard, as she stumbled and sometimes crawled up the jumble of rocks. Then, near the summit, is an especially steep part of the "trail."

"Cheer up, cheer up, Polly!" I hear [brother] George break the long silence. "Only a hundred feet to go now." One hundred feet! That sheer wall of rock! Can I make it? In some inexplicable way the men of our party get round me. They push and pull me. They turn and twist me, until my very joints creak with the pain of it. "Don't look down," they warn. I have no strength to turn my head, to speak. Only 10 feet more! Oh, God, what a relief!

Then my foot slips! I lose my balance. I fall only a few feet into a crevice in the rocks. The sharp edge of one cuts through my boot and I feel the flesh of my leg throbbing with pain. I can bear it no longer, and I sit down and do what every woman does in time of stress. I weep. "Can I help you?" "Can I help you?" asks every man who passes me. George tries to comfort me but in vain. He becomes impatient. "For God's sake, Polly, buck up and be a man! Have some style and move on!"

Was I mad? Not even allowed the comfort of tears! I bucked up all right and walked triumphantly into that broker's tent—an ancient canvas structure on the summit.[24]

Martha later discovered that she had been two-months pregnant during her strenuous crossing of Chilkoot Pass. Her third son, Lyman was born January 31, 1899 in Dawson.

The pass was still more treacherous in the early spring when soggy, melting snow slipped from the steep slopes and came raining down on surprised stampeders. Over eighty people lost their lives in just such an avalanche on April 3, 1898. Inga Sjolseth (Kolloen's) entry for that date describes incidents at her own tent at Sheep Camp which forecast the greater disaster to occur later that day.

Sun., April 3 It has snowed hard all night and continues today as well. Our tent stands right under two large trees, and the snow falls down from them in great masses right on our tent. We were awake all night. A sad accident took place here today. A great snowslide came down from the mountain about two miles from

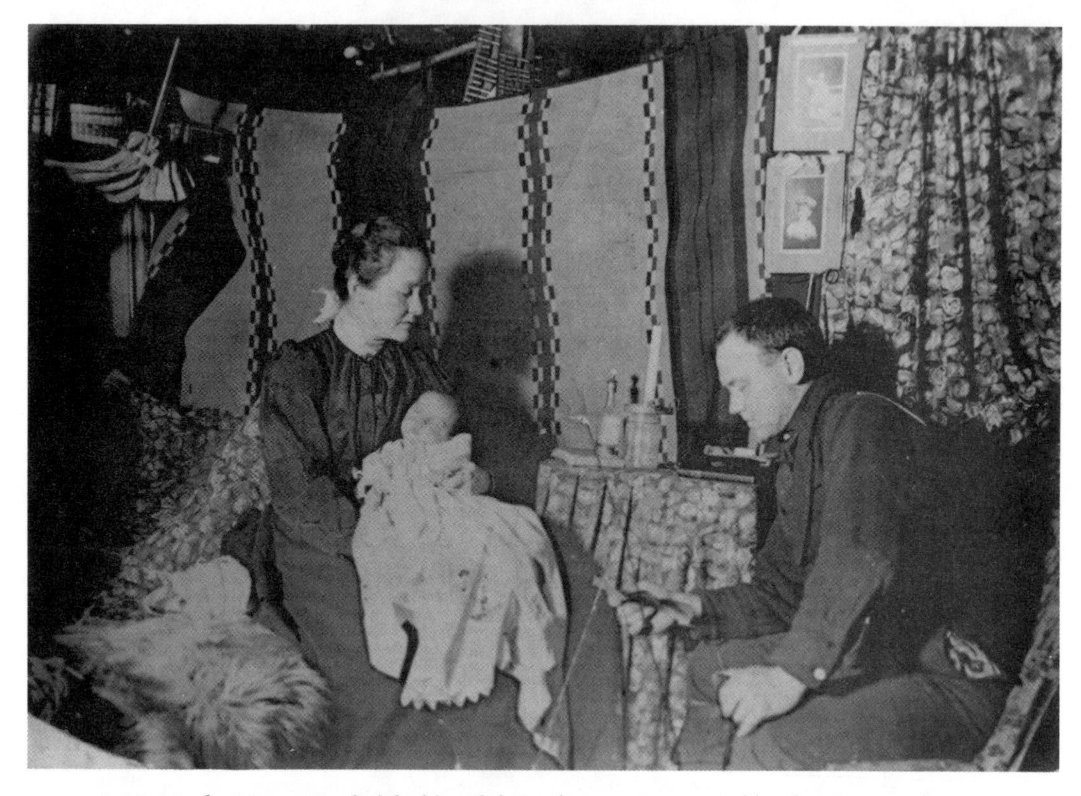

4–25. Martha Munger Purdy (Black) with her infant son Lyman and brother George, about February 1899, in the cabin they built and furnished in Dawson. (Y.A., M. L. Black Coll'n, 82/218, H–14)

> Sheep Camp and many people lost their lives. They have now dug out 39 victims. How many more lives were lost we do not know. Erickson was one of the unfortunate ones. . . . This has been a very sad and long Sunday for many people here.[25]

Edith Feero Larson also recalled her connection with that tragic episode.

> One of the folks that came up on the *Al-ki* with us in the fall of '97 lost everything in that avalanche. Jonny Caracas. . . . He was a young fella, traveling with two older men. When they got to Dyea, they unloaded everything. . . . Your raft is made of your lumber on the boat. If you just got freight, your freight is put on that, and it's cut ashore, to drift in. They just put his stuff all on the raft, the dogs and all, and they started to cut it ashore. He says, "You can't cut my dogs loose!" The captain says, "I can cut anything loose except a human being." Alright—Jonny jumped off the deck of that boat onto the raft with the dogs! "Now," he says,

> "cut me loose!" So they had to tow him ashore. So [he] lost everything, his dogs and everything in that avalanche.
>
> He come over after the avalanche and he came out to see us. . . . He told mother he was goin' back out to see if he could get another outfit. If he could get another outfit, he would be back in the . . . summer, and he would see her. We never saw him again.[26]

Tot Bush (Berry) also wrote of the Chilkoot avalanche. Fortunately C. J. had heeded the advice of his Native American packers and had insisted that his family and friends not go over the pass with the unstable snow conditions.

> Everybody rushed to the slide to help dig out those who were buried, and they were able to save some of them. A man and his wife were buried in the slide; she was taken out alive, and directed them to where her husband was buried. They had just brought him to the surface when someone yelled that another slide was coming, and the workers all had to run for their lives. The rescued man was buried again, but he wasn't found after the second slide. Two men were sleeping in a tent at the time of the first slide; the tent was cut in two and one of the men killed; the other was left asleep in his bed. All the bodies were taken to a crude morgue that was set up. All were frozen and looked as natural as in life. We helped to make coffins of plain wood, lined with black cloth; in each coffin was placed a bottle, containing the name and what little was known of the person. Had Annie [Tot's cousin] and her friends gone that morning as they had intended, they would have perished. The next day Ethel and I rode to the graveyard with five of the dead men. I am glad to say that none of our friends was killed.[27]

And yet Chilkoot Pass was not just a scene of tragedy and misery. Only one week before the deadly avalanche, a wedding ceremony was performed at the summit. The bride was Marie Isharov, a 20-year-old from Poland, who was going in with her father. She had met the groom, 30-year-old Frank Brady, a Montana miner, on the trail. And as they made their way up Dyea valley, their friendship blossomed into love, so they had decided to marry. The couple left it to Frank's friends to arrange a quiet ceremony. But the results were not so private as the couple had wanted.

> The wedding procession started from "The Scales" at 11 a.m. Leading the way, accordion in hand, was Phil Ward of Virginia City, Mont., one of the most accomplished players who ever evoked "Home, Sweet Home" from this instrument, and as the strains of Lohengrin's "Wedding March" were taken up by the

4–26. Recovering bodies after the April 3, 1898, avalanche on Chilkoot Pass. (U.W., LaRoche photog., 2130)

4–27. Stampeders hauled just about everything over Chilkoot Pass—food, clothing, mining equipment, sleds, and even a piano. Note that most have no real backpacks. (U.W., Cantwell photog., 46)

> breezes, Archie Burns tramway ceased its creaking and the workmen at Stone House stopped to listen and to look at the interesting procession going up the famous marble stairs of Chilkoot pass. . . . Following Musician Ward were Ushers Gilbert and Seigfried, also of Virginia City. Then came the bride, accompanied and assisted by Bert Fenner. Miss Isharov was sensibly attired in neat-fitting, modest Klondike garments, and appeared as happy as any bride on whom the sun ever shone. Following the bride came her father and Mrs. Decker, a handsome young widow from Puyallup, Wash., who is on her way to the interior. . . . After Mrs. Decker and the bride's father came Ushers W. A. Stevenson, Knute Ellingson and William Nurnberger, all of Virginia City, and Arvin L. Kells, of Dawson. Then followed a crowd of interested spectators, many of them with loads of 100 pounds on their backs.
>
> When the wedding procession reached the summit it was met by Gus Steffens, the well-known Dyea jeweler, and Rev. Christopher L. Mortimer, a Missouri minister who is seeking fortune and souls to save in the gold fields. . . . Resting their heavy packs on the snow, this gathering of gold seekers from all parts of the earth stood with uncovered heads and reverently watched the minister of God join in holy wedlock the handsome Montana miner and the beautiful Polish girl.
>
> . . . The wedding ring was an exceedingly heavy one, rolled by Jeweler Steffens out of gold presented by Arvin L. Kells, who himself dug the nuggets out of claim No. 27 below on Hunker creek.[28]

Chilkoot Pass, then, was a place of contrasts. It had great natural beauty—glaciers, streams, flowers, berries, picturesque rocks or snow-covered peaks. And yet it was also a source of such physical challenge and pain that many stampeders only recalled that aspect of it. Comraderie, cooperation, and optimistic good will traveled side by side with the stresses of the press of great numbers of people, competition, and exasperated tempers. One weekend the pass was the scene for a celebration of life and the beginning of a marriage, the next for mourning and the end of hopes and dreams.

Except for Native American women, most women did not carry heavy packs over Chilkoot Pass. And yet, the climb was an ordeal for them—an indication of how really trying that section of the trail could be. Other factors added to the difficulty. Like Martha Purdy (Black), many wore conventional women's clothing of the day which was inappropriate for the difficult trek—corsets, long skirts, tailored jackets, lightweight shoes. And like many of their male counterparts, some of the women were probably unconditioned to such stenuous exercise. *Harper's Weekly* correspondent, Tappan Adney, reported that the cause of trouble in getting across the trail was the "inexperience of those who are trying to go over. They come from desks and counters; they have never packed, and are not even accustomed to hard labor."[29] Even those women less articulate than Martha Purdy (Black) in describing their passage, communicated their difficulties nevertheless. As we saw, Inga Sjolseth (Kolloen) was really frightened of the pass. Perhaps the tragic avalanche that had taken so many lives in early April weighed on her thoughts. And yet she conquered her fear. Her diary says nothing of the crossing, however, save that it was accomplished April 27.

Emma Kelly, correspondent for the Kansas City *Star*, is another who crossed the pass with difficulty, though she does not provide many details. She traveled to the Klondike on her own in the late fall of 1897 and faced a blizzard getting to Sheep Camp, as she recounted later.

> I knew no person in Dyea and was unable to secure packers of experience, and facing this dilemma, I determined to try for packers among the inexperienced deck-hands of the vessels in the harbor. By promising to feed them and pay each man fifty cents a pound for all he carried over, I secured a motley crew of ten men of various shades of color and nationalities to take my goods to Lake Linderman, where I learned a party of men had built some boats and would soon start across the lakes and down the river to Dawson, my great anxiety having been to reach the lake in time to secure passage in one of the boats. . . .
>
> . . . I secured Healy and Wilson's pack train to take my goods to Sheep Camp at the foot of the pass, twelve miles from Dyea, and with my packers I fell in immediately behind the horses, and after plodding all day through slush and muck, in the face of a terrific storm of sleet which drove into our faces like tacks, we reached Sheep Camp at five o'clock in the evening. As the storm was still raging I had to lay over a day waiting for it to subside, having again to herd my packers from those at the camp [so they would not hear how dangerous the pass was]. . . .
>
> Some of my men wanted to turn back from Sheep Camp, but I told them they could make the trip if I could, and so I held them to their contracts.
>
> The following day the sun came out warm and there was but little wind. I was up at five o'clock and got breakfast, and by seven the men were packed and we all set out across the summit for Lake Linderman, eighteen miles distant. At ten o'clock we reached Stonehouse, at the foot of the steep climb, where the elevation rises for two and a half miles at an angle of forty-five de-

> grees to the summit, which is usually enveloped in mist and clouds. An immense glacier is crossed by the trail and travel is impeded by great bowlders [*sic*] which cross the narrow, winding footway.[30]
>
> I placed my packers in a line ahead of me and started for the top. We could only proceed a few steps, when we were compelled to stop and rest in order to get our breath. Thus we proceeded by slow stages to the summit. The trail, covered with snow and ice, made it dangerous and wearisome.
>
> When about a mile from the summit we met a man returning from the lakes. The foremost packer called to him and inquired how far it was to the summit. The man looked at him and answered the question with: "How many pounds have you on your back?" "One hundred." "Then it is just one hundred thousand miles to you," said the stranger.
>
> I really believe it would seem that to those who were carrying packs. When I myself asked the man the distance, he noticed I was without a pack, and smiling said: "About a mile and a half."[31]
>
> At one o'clock the summit was reached, and with a feeling of exultation I sat down with my packers for a rest and to look back over that fearful trail from Sheep Camp. I was safely over the worst part of the trip, and with all my goods.[32]

Though Emma Kelly felt exhilarated when she made it across the pass, she also felt the aftermath of the strain the next morning at her camp at Lake Lindeman. Still, it was a matter of principle for her not to complain.

> At about ten o'clock on the following morning one of the party came to my tent and told me they would start at noon. I was lying on the ground, rolled up in my blanket, and was so sore and stiff, and ached so from my long climb and walk in the water, mud and snow of the previous day, that I was unable to get up without help. One of the men helped me, and I had almost to crawl about that day, but I made no complaint that I felt badly and said nothing of my condition, for fear I would be thought a nuisance from the start [by the men from whom she had purchased passage to Dawson].[33]

At the top of the pass was the Canadian border crossing. Northwest Mounted Police (without their mounts) collected customs in a tiny tent flying the British flag, surrounded by mounds of cached goods. The cramped area between mountain peaks was a focus of activity as stampeders arrived, emptied their packs, stretched and rested momentarily, then left for another load or for the next leg of the journey into Canada.

Martha Purdy (Black) described the summit this way:

> I had made the top of the world, but "the wind that blew between the spheres" cut me like a knife. I was tired, faint, hungry, cold. I asked for a fire, and was answered, "Madame, wood is two bits a pound up here." George, who was really concerned about me, spoke up: "All right. All right. I'll be a sport. Give her a $5 fire." One heavenly hour of rest. I took off my boots, washed my wounded shin and poured iodine on it. I dried my wet stockings, had a cup of tea, and got thoroughly warm.
>
> We then went through customs, as we had now entered Canada. Around us, shivering in the cold wind, were many waiting people, their outfits partially unpacked and scattered about them in the deep snow.[34]

4–28. Stampeders pause for a photo on Chilkoot Summit on a clear day in 1898. The Canadian customs house is by the British flag, and the cable from the tram can be seen in the background. (U.W., Hegg photog., 211)

4–29. Crater Lake as seen from the Canadian side of Chilkoot Pass in the fall of 1897. The trail descends in the right foreground and follows the east shore of the lake at right. (U.W., LaRoche photog., 2046)

Summit to Lake Lindeman

On the other side of Chilkoot Pass is a long, treeless, windswept valley. Through it a chain of small lakes and waterfalls descend to the larger lakes, Lindeman and Bennett. Rocky slopes surround each lake in the summer. In the winter, the boulders and sharp rocks are covered with deep snow. Into this valley, then, the trail descends steeply to the first lake, Crater Lake.

Down the steep slope outfits could be sledded with caution in the winter. But they had to be carefully braked because a runaway sled loaded with several hundred pounds could be lethal to those downslope. In the summertime, everything still had to be packed on backs. Then new strains were felt, especially on knees, as packers fought to keep their balance with the jarring weight, unstable rocks, and glacial ice.

Without packs, however, this section of the trail could actually be fun. Tot Bush describes sledding with glee down the slope to Crater Lake. A similar account is given by Ella Hall. Ella and her sister Lizzie (Mrs. C. G.) Cheever were from Massachusetts and had been strongly smitten by gold fever. Lizzie's husband, a manufacturer and dealer for the newly perfected,

4–30. Easing heavy sleds down the Canadian side of Chilkoot Pass, April 1898. (U.A.F., C. Bunnell Coll'n, Brooks photo., 73–66–23)

4–31. Strong winds help to move goods across frozen Crater Lake. (U.W., Hegg photog., 95)

4–32. Jumble of outfits, boats, stampeders, and packers on the rocky shores of Crater Lake. (U.W., LaRoche photog., 2048)

pneumatic-tired "safety" bicycle, said that they had tried to convince him to go with them to the Klondike. But when he said he didn't care to, they went on their own, setting out from Somerville, Massachusetts, March 31, 1898.[35]

In her memoirs (apparently written several years after the event), Ella Hall recounted that she and Lizzie hired packers to carry their goods across the pass at the end of April. In contrast to many other stampeders' tales of trying experiences of the pass, Ella Hall's description contains very little about the ascent. And the descent to Crater Lake she likens to a trip to an amusement park.

> I can only compare this to a trip in Wonderland Rivere Beach, for it was needless to try to keep your footing, owing to the steepness and slipery ice. A person would roll and tumble, slide just as if on the scenic railroad or in a fire shoot. Our sides ached (with others) from laughter. The sleds had to be weighted with bags of flour and provisions, tied on behind to keep them back from going in every direction. They used the weights as braks. We then journeyed on (coasting most of the way) to Lake Linderman whare they pitched our tents, after the snow was shoveled away and they cut spruce boughs for our beds.[36]

4–33. Native American woman packs across rocks about two miles beyond Chilkoot Summit, 1897. (A.M.H.A., Fridley coll'n, B70.22.44)

Ella Hall's experiences can be contrasted to the hardships encountered by Martha Purdy (Black). She covered the same fifteen miles between Chilkoot Pass and Lake Lindeman, but a couple of months later in the summer when the trail was along boulder-strewn lake shores.

Even though Martha had thought she was over the worst part when she reached Chilkoot summit, after supper she found that Chilkoot trail still held many challenges.

> Then the descent! Down ever downward. Weight of body on shaky legs, weight growing heavier, and legs shakier. Sharp rocks to scratch our clutching hands. Snake-like roots to trip our stumbling feet.
>
> We stopped at the half-way cabin for a $2 supper of bean soup, ham and eggs (of uncertain age), prunes, bread, and butter—the bread served with the apology of the proprietor, "The middle of it ain't done, but you don't have to eat it. I hurried too much."
>
> I had felt that I could make no greater effort in my life than the last part of the upward climb, but the last two miles into Lindeman was the most excruciating struggle of the whole trip. In my

4–34. Interior of typical tent restaurant along Chilkoot trail on September 4, 1898. This one was at Crater Lake, just beyond the summit where Martha Purdy (Black) stopped for refreshment. (U.W., Hegg photog., 3100)

memory it will ever remain a hideous nightmare. The trail led through a scrub pine forest where we tripped over bare roots of trees that curled over and around rocks and boulders like great devilfishes. Rocks! Rocks! Rocks! Tearing boots to pieces. Hands bleeding with scratches. I can bear it no longer. In my agony I beg the men to leave me—to let me lie in my tracks, and stay for the night.

My brother put his arm around me and carried me most of the last mile. Captain Spencer hurried into the village, to the Tacoma Hotel, to get a bed for me. It wasn't much of a bed either—a canvas stretched on four logs, with a straw shakedown, yet the downiest couch in the world or the softest bed in a king's palace could not have made a better resting-place for me.

As my senses slipped away into the unconsciousness of that deep sleep of exhaustion, there surged through me a thrill of satisfaction. I had actually walked over the Chilkoot Pass! . . . I would never do it again, knowing now what it meant. . . . Not for

4–35. A party of missionaries in camp on the rocks at Long Lake, just above Lake Lindeman. The women sport shorter skirts, more suited to hiking than those that were fashionable. Edith Larson said, "The husbands and wives worked together (on the trail). The wife, she went over the trail, she put on overalls or pants like a man, 'cause she couldn't travel in skirts. You know, in those days, skirts was down to the ankles. Well, you'd have a heck of a time goin' over the trail in that!"[38] Even so, photographs of women on the trail often show them with long skirts. (U.W., LaRoche photog., 2050, p. 21)

all the gold in the Klondyke. . . . And yet, knowing now what it meant, would I miss it? . . . No, never! . . . Not even for all the gold in the world![37]

Stresses and Good Will

Surprisingly few of the Klondike stampeders had had much experience in the wilderness, and practically none had been to the Arctic environs before. So Martha's feelings about her crossing of Chilkoot Pass and the subsequent hike to Lindeman were not unusual. Many stampeders turned back at this stage, content with having simply seen Alaska and congratulating themselves for having the sense to return home while still in one piece. Those who continued to Lindeman felt a sense of accomplishment while, like Martha Purdy (Black), vowing to themselves never to do it again. But even for these brave ones who continued, the physical dangers, illness, exhaustion, hunger, cold, disappointments, frustrations, and tragic losses of life were sometimes nearly unbearable. Many reacted to the strains by directing their frustrations outward, and numerous stories are told of partnerships that ended on the shores of Lakes Lindeman and Bennett. Here is one from Edith Larson.

> I met the man [who argued with his partner] all the way over [the pass]. Then when they got to Bennett, they got mad, and they split their outfit. One took the stove and one took the tent. One took half the provisions, the other. . . . They cut the boat in two so that each had half! Well, that night it rained like the dickens. The one in the tent was dry, but he was cold. The one with the stove, he had the stove but he was wet. The next morning they got up, patched the boat up, made friends, and went to Dawson in that boat![39]

Other partnerships came to more permanent endings. For example, after almost two months of tension, Inga Sjolseth (Kolloen) describes the breakup of her party.

> *Tues., May 17* Conflicts are now the order of the day here. Ida is very angry with me and finds fault with me all the time. She accuses me of things I have never done, and it is more than I can stand. I contradict her and we get into real battles.
>
> *Wed., May 18* Ida was almost beside herself today. She took down our names from the tent and cut off her name. She says that Hansen has made fools of us. She says I am so bad that we cannot longer live together. I shall not treat her like a little sister any more, for it is impossible for us to live in peace.
>
> *Thurs., May 19* Mr. Anderson and Miss Bodin [Ida] have moved from us today. They set up their tent a few miles from us and are now planning to live together there. Anderson and Sandvig have divided all the supplies they owned together, so now they have become two separate units.[40]

For others the stresses were turned inward. Georgia Hacker White, for example, barely had been making a living at home. She decided that the Klondike offered better prospects. So she left her three children in the care of friends in Reno and the Ladies Protective Relief Society in San Francisco, and set out February 21, 1898. Unlike many of the women we have seen, she seems more motivated by desperation than by the spirit of adventure. Throughout her diary of the trip are notes that she is not feeling-well, or that she is very homesick. At Bennett she goes "to the English

4–36. Georgia Hacker White and a Mr. Flynn at Dyea in the spring of 1898. (Dorothy DeBoer Coll'n)

officials to see poor Emma who is insane. She looks dreadful and did not know us."[41] A few days later on June 27, Georgia sounds depressed and worried about her own sanity.

> I fear very much that I am a damper on our company, I am so quiet and seldom if ever make fun or seem good-natured. Firstly, I think constantly of my little ones and God knows at times it seems more than I can bear but I must—for Oh deliver me from becoming insane up here. I wished I could cut up more and make it pleasant for those around me for I feel so kindly toward them all.[42]

For some the stresses were simply too much. Lake Lindeman is the first place after the pass where trees grow. Although timber at Bennett, the next lower lake, was larger, stampeders often built their boats or rafts at Lindeman, eager to begin the long journey by water to Dawson. Those who did and waited for the ice to break up, however, had to navigate the short but treacherous rapids between Lindeman and Bennett.

4–37. One boat shoots the One-Mile (King River Falls) Rapids between Lakes Lindeman and Bennett in 1898, skirting an already-wrecked raft. In the distance at center is Lake Bennett. (U.W., Hegg photog., 2115b)

4–38. Grave above rapids between Lakes Lindeman and Bennett of a discouraged stampeder, J. V. Matthews, who shot himself after losing two outfits there. (This may be the A. Mathews reported in some newspaper articles to have committed suicide, and who was former sheriff of Puyallup, Washington.) (U.W., Larson coll'n, 9083)

Edith Larson tells of one stampeder whose "Yukon test" came at these rapids. "King River Falls joins Lake Lindeman and Lake Bennett. . . . Up on top of the canyon there is an old grave, and it was from a man that had lost two outfits in King River Falls. And when he lost his second one, he went up there and committed suicide. They buried him just where he committed suicide."[43]

It is probably significant that this despondent stampeder seems to have been traveling alone. Companionship on the trail often proved an important element in emotional as well as physical survival.

IF THE Dyea-Chilkoot trail was rigorous for adults, it could be especially difficult for the children who accompanied them. Not only were there the hazards of physical strain and sometimes life-threatening weather, but also diseases—hepatitis, meningitis, typhoid—were passed between stampeders in the crowded, makeshift camps along the trail. Florence Hartshorn, who lived at Log Cabin on the nearby White Pass trail during the gold rush, recalled that at least three children were buried on the Chilkoot trail—one at Dyea and two at Lindeman.

> Mr. and Mrs. McKay were going into the Klondike. Mr. McKay was an Alaskan trader. Telling of her trip [Mrs. McKay] said it was grand only for some sadness, for one morning . . . while she and husband were at Lake Lindeman, a Mr. and Mrs. Card and child were also going inside and camping nearby. Mr. Card came to their tent saying their little one was dead. This was their first child of 7 months. She said, "We showed the parents our sympathy by helping all we could. We made a small coffin of rough wood padded with a soft blanket, covered the outside with black cloth, lining it with white cloth." Mrs. McKay said, "I laid the baby in it, placing a bunch of violets from a hat in its little hands. In a small tent the little one laid until the next day, when with loving hands the child was buried. A wooden stake at the head told the travelers who was lying there. A picket fence was built to protect it.[44]

Who, in selecting a hat decorated with artificial flowers to take into the wilds of the Yukon, could have anticipated that it would provide the only symbol of life for an infant's coffin?

But the child's parents went on. The Fred Cards had lived in Juneau since 1895, so perhaps their familiarity with the north country had given them the confidence to set out with their infant daughter on the winter trail. (Ella, Fred, and infant daughter are probably pictured in Figure 4–9.)

4–39. Two stampeders in camp near Lake Bennett. (A.H.L., Metcalf photog., PCA 34–91)

One can only guess at the regrets Mrs. Ella Card faced as she and her husband continued to Dawson in the spring of 1897.

Though Florence Hartshorn does not say so in her notes, the McKays who helped the Cards may also have lost their child near Lindeman. A Mr. and Mrs. J. D. McKay came across the trail in 1897 and are reported to have buried their infant daughter next to the grave site of the Cards' child.[45]

As in the story above, stampeders' diaries and letters often tell of mutual responsibility and goodwill shown in time of need—not that all were generous or good Samaritans, but many were, despite the competition to get gold. Women, perhaps more than men, were expected, by themselves and others, to act selflessly and to respond to others' needs. In this account from Florence Hartshorn's notes, we see how these expectations could go well beyond the reasonable. Kate Ryne, a young girl on her own, arrived at Telegraph Creek, on the Stikine trail, shortly after spinal meningitis broke out. Kate, who was not a nurse, nevertheless took care of the sick as best she could. And, as Hartshorn recorded, when they died, she laid them out using whatever materials could be found.

> Someone died. The coffin was made of rough boards. She took some cloth . . . ,placing leaves and twigs. On outside was covered with cloth from an old black skirt. Now her trouble was to get help as the present body was too heavy for her to lift. So she asked for help. At last a man was found who came. He put his hands on the bedding and said, "This is not soft enough." . . . The men had to get the [bedding] full before they would touch one of the coffins. While she found a few more boughs and leaves; used his hand to test the softness, saying, it wasn't yet soft enough. For said, "This old prospector is entitled to a soft bed. He's gone through hell."[46]

Later Kate Ryne highlighted the ambivalence she felt in complying with the miners' demands by explaining that she filled the coffin so full of leaves that when they tried to put the body in the box, it rolled out! The corpse let out a groan as it rolled, so frightening the man who was "helping" her that he ran. The "deceased" was, in fact, dead; apparently air had been forced from his lungs by the change in pressure on his chest. Despite exposing herself to the deadly meningitis to nurse those who were ill and lay out those who died, Kate Ryne never charged for her services but told all to give to someone else who was in need.

Kate Ryne never did make it to Dawson.[47] But many thousands of other stampeders did continue—despite all the hardships, disease, and physical and mental stress.

SKAGWAY, GOLD RUSH BOOMTOWN

"I always wanted to try making my own way"

THE TRAIL that originated at Skagway and crossed White Pass was the brainchild that Captain William Moore conceived for the gold rush. It did not exist before 1897. Ten years earlier Moore homesteaded what would become the townsite of Skagway. Before then, coastal natives, probably Tlingits, had camped and hunted in the area.[1] Located at the most northern part of southeast Alaska, Skagway faces Lynn Canal to the southwest. Steep, usually snow-covered mountains flank it in all other directions. The Skagway River, which empties the watershed to the north, flows along the west side of town to Skagway's wide, sandy beach.

Captain Moore had the foresight to expect that eventually many people would want to get to the mineral wealth of the interior without having to contend with Chilkoot Pass; he had recognized that White Pass summit was nearly 750 feet lower than the neighboring Chilkoot. And then, his timing was good enough that he had already laid out the approximate best route for a trail to White Pass before the stampede began in July of 1897. When boats loaded with gold seekers began to arrive in Lynn Canal, they were actually induced to land their passengers at Skagway to use the new trail rather than the one from Dyea. Within one month, Skagway grew from a one-cabin homestead to a bustling boomtown.[2]

Ten-year-old Edith Feero's arrival at Skagway in the late summer of 1897 was full of confusion and excitement. As at Dyea, no pier was built yet at the "Gateway to the Klondike." Everything was landed in the shallows offshore at high tide—people; horses; dogs; donkeys; mules; cattle; sleds; boats; and boxes, trunks, bags, and sacks filled with goods of all sorts—to wade or be hauled from the wet beach at low tide.

5–1. Captain William Moore's homestead at the mouth of the Skagway River before the gold rush. This view is from the harbor looking northeast (U.W., Larson coll'n, 9078)

5–2. The same general area of the beach shown in Fig. 5–1 but looking more northward and showing the tent town which sprang up in the late summer of 1897. (U.W., Sarvant photog., 3)

When we got to Skagway, the harbor was full, so they couldn't anchor out there. So we went on into Dyea and unloaded at Dyea first and then came back the next day and landed in Skagway. But while we was coming over from Dyea, the Captain came and took us children up and said, "I want you children to have fruit and candy when [you] get off. I don't know if there's any there or not." And he gave us a lot of fruits and nuts and candy to carry off the boat. It was real thoughtful. And we had a good time!

[When] we got into Skagway there was no dock. . . . The beach was all full of scows and lighters fixed for the landing. Captain went down to the deck and he said, "All you men and boys, put on your boots. We'll carry the women and children, but you men have got to wade." And that's exactly what they did. They'd take 'em ashore as far as the lifeboat would go, then the men and boys got off and walked. But all the children and women they carried ashore.[3]

On shore was a bewildering array of activity. Stampeders trudged to and from the water line in tall rubber boots, moving recently landed goods to higher ground and the watchful guard of a partner. Horses, some only half-tamed, stood tied to stakes among the tents lining the shore. Occasionally one would bolt in fright at some unusual sound, knocking over tents, supplies, and all else in its path, accompanied by a chorus of howling and barking dogs. Everywhere there was movement.

When Edith and her family landed on the beach in the midst of all the confusion, her father, John Feero, was not there. Because of a communication mixup he had not realized that they were going to be on the *Al-ki.*

He didn't know we was on that boat. Mother had written, "We'll be on the *City of Seattle* or the next boat following. Well, he thought she meant the next *City of Seattle.* . . . Mother had 25 cents in her pocket—with four children! Well, you know, the Captain didn't trust her [to get] very far with that. He said, "I'm going to go up to the steamboat office and see if they can find Mr. Feero in town."

. . . He took her up to a tent on the beach and [they] went in. It was the steamboat office. He said, "Does anybody know Mr. Feero here in town?" Somebody out on the street yelled, "Who do you want? Sandy?" [John had a red mustache that earned him that nickname.] Well, mother was mad then. She flew to the door and . . . said, "Who are you . . . what are you callin' up here?" "Well," he said, "you know we Tacoma people. . . ." He was a man who had lived a block and a half from us in Tacoma!

5–3. Lighters and horse-drawn wagons land people and goods on the beach at Skagway. Since most of the town is still in tents, the photo was probably taken in early August 1897. (U.W., Larson coll'n, Winter and Pond photog.; also A.H.L., PCA 87–653)

5–4. A slightly different view of Skagway beach looking more eastward, probably in the fall of 1897. More wooden buildings are now evident, and two wharves are near completion, though lighters and wagons are still being used. (U.W., A. Curtis photog., 46104)

He took us up to a little small cabin he had and left mother, sister, brother [and me] there, and he took the older boy and went up to where father had his horses. . . . [Father] came down and he said, "I'm going to take you . . . to the hotel for dinner." . . . [The hotel was] a big building built with boards up and down, like they used to build a barn. [We] went in and there was a long table on one side of the room . . . with benches on either side. On the other side of the room there was a long bench and nails in the wall. That was the coat hanger. And mother looked around, . . . "I can't see no place . . . to sleep." [But] that was the hotel.

. . . We wanted to know if we could go up and see where [Father] kept the horses; whether he would let us stay there or down in town. Went up there . . . [and Mother] decided to stay up where the horses were. [But Father said,] "We'll have to stay at the hotel tonight," so [we come] back down. [Mother] rubbered around the room. "Still can't see no place to sleep." "Well," [Father] says, "you will. . . . You see that ladder over there?" "Huh! I can't walk up that ladder!" She was a state-of-Mainer, my mother. . . . She was born and raised in the state of Maine and [there] everything was proper. Well, she found out that she was either going up that ladder, or she wasn't going to sleep. So she finally got up the ladder. . . . Your skirts hang down when you're going up. They were pretty good going up.

Got up on top and looked around. Nothing but canvas curtains hanging; divided the beds. There were mattresses on the floor and canvas curtains to divide them off. We had three single mattresses for the six of us. . . . Well, we got along on 'em. We had a ball, us kids. We thought it was just loads of fun. . . . But not Mother. Mother, she didn't sleep. Course there was a lot of drinking, and there was a drunk on this side of the curtain, and a drunk on that side of the canvas, and she couldn't sleep. She got up in the morning and . . . took a rubber down the hole. "I can't go down there. There's a lot of men sittin' down there." Well, she found out she was either going to go down there, or she wouldn't eat either. So she got down, and she stomped her foot and says, "I'll not go up there again. I'll live in a tent first." Into a tent we went![4]

Although there were a few wooden buildings, like the hotel, in Skagway in late August of 1897, most people and businesses were still in tents. The muddy paths that led through town were lined with canvas saloons, dance halls, groceries, outfitters, and shops of all sorts. As Edith described it, the Feero family's tent and its furnishings would have been similar to those set up as "hotels" to accommodate stampeders. For those who slept in their own tents before striking out on the trail, the Feero family abode would have seemed elaborate.

5–5. The east beach of Skagway in the fall of 1897 when new buildings appeared among the clusters of tents. Captain Moore's sawmill, in the foreground, supplied most of the lumber for construction. Largest building to left of center is the Burkhard House, a hotel, its roof still incomplete. It was located on the corner of Broadway and Fifth. (U.W., Alaska cities, Skagway #7, 8147)

5–6. Stampeders, dogs, and tents in Skagway in late fall of 1897. (U.W., Larson coll'n)

5–7. Some of the tent homes in Skagway in the fall of 1897. (U.W., LaRoche, 9077)

Put the tent up where the horses were. The tent was in what was later the middle of Main Street, but there was only just trails around then. And we got the tent fixed up . . . [Mother had] brought three mattresses and springs. They put two of them on boxes . . . for the children, to keep us up off of the ground. The other mattress was put . . . [on] posts in the . . . ground. And that was put up high, so we'd have storage underneath. That was where mother and father slept. In one corner we had a little stove mother brought, with a drum and a chimney. That was the oven . . . She didn't have no table, so father decided he'd go to buy some lumber and build a table. Couldn't buy some lumber. So the only thing he could get was an old packing box, . . . but he had to rent it so he couldn't tear it apart. They drove posts in the ground and put the packing box upside down over that. And so we had a table. Now you got your bed, your table, and your stove. You ain't got no place to sit! So they goes back to the woods again, cuts some blocks off, chair size, and then we had chairs and everything.

The floor was all wet. It was fall of the year, you know, so it was all wet. So they out to the woods again and got fur boughs and covered the floors . . . [to] keep us kids' feet out of the water. That was our home when we first landed there. We lived in there

5–8. Skagway as laid out and ratified by the then-residents in mid-August, 1897. Broadway and Main are two of the major streets running the length of the valley, while the numbered avenues run more or less parallel to the beach. (Dedman's Photo Shop, Skagway).

5–9. The Feero cabin on the outskirts of Skagway and a loaded pack train. By 1898 when this photo was taken, Feero was in partnership with Pells. Just to the left of the Pells and Feero sign are Emma, Edith, and Ethel. By the door of the cabin is Emma's cousin Annie. John stands with his sons Willie and Frank among the horses. (Edith F. Larson)

> until the day before Thanksgiving of 1897, when we moved into the cabin my father had built on his lot.[5]

John Feero had gone north with the intention of prospecting for gold in the Klondike. But once at Skagway, he didn't have enough money to go on into the Yukon. So he stayed on the coast, worked for a packer until he had earned enough to buy horses of his own. Then he too ran a pack train. Business was profitable, so the Feero family's conditions improved rapidly, paralleling the equally booming rise of the town. The Feero log cabin was soon under construction on the corner of Nineteenth and Main Streets, though that is not exactly where they had thought it would be.

> My father had the second lot back from Main. Somebody come in and started a cabin on the first one. Then mother was tickled. "We're going to have company." Well, we got up in the morning and here's a great big saloon sign on the roof. . . . So Father had to buy the cabin to get the saloon out. That's how we got clear over on the corner. And we had those two [lots]. . . . They put a long

> tent in behind the cabin . . . for horses. Tents would cover 100 horses when he got done with it![6]

By the fall of 1897 not only were cabins built throughout town but streets were laid out and lined with stores, hotels, casinos, bars, brothels, dance halls, a church, and small businesses of all sorts. One early visitor remarked:

> The houses are mostly small frame structures of the packing-case pattern with about the same amount of finish, and the chief architectural feature of the main business street is a perfect forest of signs which stick out from the fronts of the buildings on either side of the street so densely that, looking up the street from one end of it, the fronts of the buildings are invisible.[7]

A weekly newspaper, the Skagway *News,* was started October 15, 1897. One early edition carried Annie Hall Strong's account of her joining the stampede and arriving in Skagway. Of particular interest for boosters of Skagway and the White Pass trail was her evidence that Skagway did not deserve its unsavory reputation as a lawless town.

> There are two sides to the life of the woman pioneer. One represents hardships and privations—hope deferred, which maketh the heart sick, trials and disappointments. The other presents fuel for the spirit of adventure, and its attendant excitement, leading one on and on in the hope—in this case—of golden reward, and the fondest fruition of one's most cherished dreams.
>
> When the most contagious of all fevers, known as the "gold fever," began to rage I was among the first to contract "acute Klondicitis" and immediately started northward to the land of "plenty of gold."
>
> The awful tales of suffering and privation counted for naught as long as shining nuggets were to be the reward. All went well until we reached Juneau, and here my heart almost failed me; for judging from the terrible tales told by returning and disheartened gold seekers, I was to fall among the riff-raff of the whole country at a place called Skaguay, which was the initial point of the then congested White pass. Cut-throats and mobs of evil-doers were said to form the population, and it was alleged that they lay in wait for the arrival of "tenderfeet." Was it any wonder I hoped the time would be long ere we reached the awful place?
>
> However, on the morning of the 26th of August we steamed around a point into a bay and right before us lay the really beautifully situated little tented town. It looked peaceful enough from the deck of the steamship Queen, but the faces of the future El-

dorado kings looked anxious and for once I remained behind, while the gentlemen of our party went ashore to find a camping place, thinking I would just as soon postpone my entrance into this modern Sodom until it became compulsory.

Towards evening, with fear and dread, I actually ventured ashore. To my surprise I found a surging crowd of people busy as bees rushing hither and thither—but everything was olderly and quiet. No one attempted to rob or mob. Everyone was kind, and those that were already settled assisted in every possible way to smooth over the rough places and brighten camp life for the argonauts.

There appeared to be a general feeling of bonhomie between these friends of a day. Kindness was the watchword; there was no evidence of violence or crime—nothing but kindness.

Such were the people of this much maligned town the day I landed, and there has been no change, and, in saying this, I think I voice the experience of every woman found within the confines of Skaguay.

But the little tented town is a thing of the past, and in its stead has sprung up a bustling town of 3,000 souls, and we old settlers, that have grown up with the place during the past three months, are proud of our town. Skaguay is the baby city of the world in age, but come and look at her and be amazed!

We have a church and school house. We boast of electric lights, a telephone system and other adjuncts of modern civilization, and when our tramway and wagon road shall have been completed, Skaguay will be the principal gateway to the interior. Brave, staunch and upright citizens built Skaguay, not riff-raff, hence our prosperity.[8]

This rather optimistic account of Skagway's beginnings fails to acknowledge that Captain William Moore's homestead was simply overrun by these "brave, staunch and upright citizens," who threw him off his land. Only after a four-year court battle was he in any way compensated for the claim jumping.[9] And there is no mention in Annie Strong's story either of the con men and gamblers who did indeed lay in wait to exploit naive stampeders.

By November of 1897 wharves, including Captain Moore's, were completed on Skagway's waterfront, as were telephone lines to Dyea. To accommodate the increasing number of children in town, a school was organized. Skagway's childhood pioneer, Edith Feero (Larson) remembered that it met in the newly constructed Union Church, the first in Skagway.

We went to school, after a fashion. The school was anywhere they could hold a class. They built a church. The Reverend [Rob-

5–10. Skagway Main Street, October 1897. Broadway, two streets to the east, was actually the main street in town, however. (U.W., Hegg photog., 36)

> ert] Dickey was the leader. . . . He built the church and each denomination had an hour for Sunday. . . . It was a church on Sunday, it was a meeting place at nights, it was a school in the daytime. All in one building! The preacher lived in a little lean-to in the back of the church. . . . The teacher was anybody they could find could teach a class. And each parent would put in so much money to hire that teacher. So . . . he or she would stay until they got enough money to go on In. Well, then you're out [of a teacher] again, and you pick up somebody else.[10]

Edith recalled that Skagway's early teachers tended to stay only a short time before continuing north to the gold fields. One of these was Mrs. V. A. Longuet, who went on to Dawson via the Yukon River route and wrote of her experience in *My Trip to Alaska in '98.*

Like many twins, Edith and her sister Ethel had quite different temperaments. Ethel was always getting into fights, while Edith explained, "If I could get out of a fight, I'd get out of it. I don't think there's anything in

5–11. Constructing the first church in Skagway, November 1897. Rev. Robert Dickey, who organized the project, is fourth from the left (counting the child). The church building was dedicated December 12, 1897. (U.W., Larson coll'n, Hegg photog. (?), 9082)

5–12. The first Skagway school class, June 24, 1898, which met in the Union Church. Edith is probably standing at the right end of the first row. Mrs. V. A. Longuet, the teacher, is at the left, and Dr. Campbell, the Episcopalian rector, is at right. (U.W., Larson coll'n)

fighting, and it's hard on the eyes!"[11] One consequence was that Edith often had to stay after school.

> [The teacher couldn't tell] me from Ethel, and so if she had to keep Ethel after school, she kept us both! And mother said, "Why don't you let Edith come home?" "Huh!" she said. "Because I can't tell 'em apart, and Ethel might get out and give Mabel another lickin'!" . . . Poor Mabel. She lived further up the road than we did, so she had to pass the house. So I had to stay after school.[12]

Edith's life in Skagway that first winter was certainly different than the mild winters she had been used to in Tacoma. Although Skagway is protected by mountains on either side of Lynn Canal and its climate is moderated by the water, it has plenty of snow and below-freezing temperatures. And the winter winds, gathering force as they are funneled up Lynn Canal, can pack a real whallop by the time they reach the beach at Skagway. It was this, in fact, which gave Skagway its name, for "Skagua" was an Indian word which referred to a very windy place, one where the same air is never breathed twice.[13] "[Skagway] was in the woods, you know, and the trail rode down and around. . . . We kids weren't very big, . . . and the wind blowed so hard that when we come out to the river bank, the wind would get a clean sweep and pick us right up and flop us right down on the ground! It was really rough!"[14]

The Christmas of 1897 was a happy occasion for the Feeros. They had a snug, warm cabin and enough money to live comfortably. Prospects for the new year looked even brighter. The Christmas season also produced some surprises.

> In early days, they used to set the table, . . . then you covered it with a big cover. Grandmother made us a cover for the table, and she embroidered a little mouse in one place. And she said, "That's my place at the table. I can't come because I get so seasick." She couldn't even go on a boat tied up at the dock! "But that's my place at the table. I'll always have a place at your table." So when the table was set, Grandmother's place was put there too.
>
> So this Christmas we had our Christmas dinner and had just about finished dinner when somebody raps at the door. My Dad went up to the door, and this fella stood there. . . . He walked in, and saw this place at the table hadn't been used, so he sat down and ate his dinner! When he got all through dinner, he said, "Well, I suppose you folks are wantin' to know why I came for. I came to tell you we're having a Christmas tree downtown, and we want you to bring your children and come down. There's a law

5–13. Christmas celebration at the Union Church in Skagway, probably 1897 or 1898. (A.H.L., Sincic coll'n, PCA 75–71)

> against vile shirts or neckties. You'll be fined if you wear one of those! But bring the children and go down." So we did. Turned out to be the garbage man, when we found out who it was![15]

That Christmas party was the first of many such community celebrations. It was held in the newly built, all-purpose Union Church. As was typical of the early days, entertainment was spontaneous and "homegrown."

> Went to the church, and the church program was just anybody. If you was there and could sing a song, well you'd get up and sing a song; if you had a little piece, then . . . , just anything you could do. That was our entertainment. We had the Christmas tree. The Christmas . . . program was about half over when Willie got sick!
>
> Mother said, "You stay with the children. I'll take Willie home. He's a big boy now, and I won't be afraid to go home." So they start out, and pretty soon in she comes, plowing in in a hurry! And she says, "I'm not going home! There's some drunks up there!" . . . They came out of the saloon as she was passing by, and "Why-ya whoop!" and fell off the sidewalk! She turned and

run. She was scared to death. But you know, Willie wasn't sick any more that night.[16]

Many stories are told of the bravery of the stampeders who went over the trail to Dawson. But those who stayed and built the towns along the way had opportunities to be brave as well. For bravery is learning to face one's fears, wherever they appear, no matter if through someone else's eyes the situation may appear harmless. For example, Edith's mother, Emma, apparently felt quite threatened by unseen, potential attackers during most of her life in Alaska. And yet she followed her husband to Skagway, reared her children, and lived the rest of her life there. "My mother used to be [frightened about entering the house after she'd been away.] Take her home, you had to go through everything. Look under anything a person could get under, 'cause she was scared. Always was [that way]. 'Course, where she was raised back in the state of Maine, [there was] no drinking, no wild life. I always figured that's what made her that way."[17]

The children of Skagway apparently shared none of Emma's fears. From Edith's description of her childhood, a typical summer or weekend

5–14 An early photo of children exploring one of Skagway's boardwalks. (Y.A., MacBride Museum Coll'n, H. D. Banks photog., 3986)

day included groups of kids exploring the town and its environment at will as they played. Most entertainment in those early days was invented, children's games as well as community celebrations.

> Before they had television and before they had radio, people used to make a lot of their fun. Our first pack of dominoes . . . Mother made it by cuttin' card board and putting the figures on. That was our first game of dominoes. We never had a pack of cards in our house—weren't allowed in the house—until Father came home one day. "I brought a pack of cards," he said. "I thought there's no harm in the children playing cards here if we just play games." But there's no money on it.
>
> . . . When we didn't know what else to do and the weather was good, the neighbors would gather together and we'd go over to the river bed and play "Run, sheep, run!" . . . You get on two sides, see. You draw sides, half to one, half to the other. Then one is a leader on each side. And they take you out and hide you somewhere. The other outfit is blindfolded, or else they stand in the corner where they can't see. And they hide this one outfit. . . . Then the other outfit goes out and tries to find 'em. And if they get too close, you'll hear somebody, "Down!" If they get too awfully close, then they yell, "Run, sheep, run!" And then they scatter and run like the dickens and hide again. And we'd play that half the night.
>
> We'd go to different neighbors' houses, and they had mouth organs. They didn't have piano everywhere then. They had a mouth organ, a violin, some of them had these jew's harps. . . . And we'd throw the rug back, if they had a rug, and fill the floor with corn starch. Just cover it up with corn starch. Because you know, if your floor is rough and you keep starch on it, they dance on the corn starch and it fills the cracks up. And after a few times, your floor's pretty flat. Then we'd have a dance.
>
> . . . And then they had a—I don't know how they done it. I tried to figure it out the other day. You make with your hands. And sometimes you make different figures to shadow on the wall, between the light and the wall. And you [formed] shadow pictures.[18]

Skagway grew quickly, and for the ten-year-old Edith, it was full of new experiences and excitement. Thousands of people from all over the world landed at Skagway or Dyea on their way to the gold of the Klondike. Many of them found they could make a good business of serving the needs of the travelers and so abandoned their trek and set up businesses and homes. In the last six months of 1897, Skagway went from a single home-

5–15 Skagway in January 1898, looking west across the valley at about Fourth Avenue. The beach is in the middle ground at far left. (U.W., Hegg photog., 455)

stead to a town of a few thousand "permanent" residents. And ships and boats kept landing, two or three a day, filled to capacity with ardent gold seekers.

Among the stampeders were also a few with a goal other than gold. One of these was the seventy-six-year-old woman interviewed by Skagway newsman, Stroller (E. J.) White. Barbara (last name unknown) came to Skagway in the spring of 1898 just to see what the rush was about and to take up a challenge. She was a tiny, gray-haired, blue-eyed widow from Butte, Montana, who had been visiting a friend in Seattle when she decided to take the money her daughter had sent her for passage home and go to Skagway instead. "All my life I've wondered what it would be like to go out among complete strangers and make my own way," she explained wistfully. "I always wanted to try it, and never had the chance. When the chance came, I took it. And here I am!"[19] She convinced the Stroller to give her a job selling papers. The newsman admitted he didn't expect her to last long, but Barbara surprised him.

> Barbara began selling papers and the first money she earned went for a place to live. She paid two dollars for a piano box and made a home of it, a merchant kindly allowing her to locate her home on a vacant strip of ground next to his store. Of course, she took her meals out, but the piano box served as a place to sleep—

> if Barbara ever slept, which the Stroller sometimes doubted. It was not long before Barbara was selling more papers and making more money than any dozen paper-sellers in town.[20]

Barbara was a hustler. She soon made friends throughout town who became her loyal customers, and she met every steamer, ready to supply its passengers with the news they so eagerly wanted. According to the Stroller, Barbara also had one sales method all her own.

> A big rawboned fellow [in a saloon] had just lost seven straight times on the black and was giving vent to his feelings. He cursed the game, the dealer, the country, the climate and several other things that came to his mind, and it was plain to see that he was no green hand at the business of expressing himself. As this molten flow poured forth, Barbara entered the place and walked directly to the fellow. She stopped in front of him, turned those blue eyes up to his face and with that appealing smile, she asked: "Would you like to buy a paper mister?"
>
> The cursing stopped abruptly and the fellow stood there with his mouth wide open. His face turned red. Then he shut his mouth and thrust a hairy paw into a pocket. He pulled out a silver dollar [papers cost ten cents], thrust it at Barbara with a muttered, "Keep the change," grabbed a paper and buried his face in it as though it were the most interesting thing he had ever run across. Barbara thanked him with another smile and went on to the next customer.[21]

Barbara saved all her money except what she needed for meals. By the fall, she had $1,350, and with that she went back to Butte for the winter. She wrote to the Stroller asking whether she could have her job back in the spring and requesting that her piano box home not be disturbed. According to the Stroller, though Barbara's piano box remained standing next to the store on Skagway's main street for several years, she never did return.[22]

But Barbara had made her own way at least once in her life.

PACKING ON THE WHITE PASS TRAIL

"I can be a lady on the trail"

CAPTAIN MOORE'S Skagway/White Pass Trail had a very different character from the nearby Dyea/Chilkoot route. Instead of people being the main source of transportation across the pass, pack animals—horses, mules, burros, and even cattle—could be used the entire distance to Lake Bennett. As a result, valuable time was not lost switching goods to smaller backpacks, as was necessary to cross rugged Chilkoot Pass. To many this was a distinct advantage and one for which they were willing to pay, if they had the cash. They either brought pack animals with them or hired a professional pack train at Skagway. Others reasoned that if the trail were better for pack animals, it would also be better for humans, so they decided to trek in via White Pass, despite the fact that the trail was longer than Dyea/Chilkoot.

But the Skagway trail had severe drawbacks, as stampeders soon discovered. It was not really completely laid out, and the route which had been selected had not withstood the test of time, as had the Chilkoot trail. In fact, Captain Moore's trail was in such poor condition by late summer of 1897 that an ad hoc committee of stampeders actually blockaded the way and required everyone to help improve it. Some bridges were built. And a corduroy road of logs helped to cover some of the areas which had been churned into mud pits by the thousands of stampeders who had already passed. Tappan Adney, *Harper's Weekly* correspondent, described one such section as "a continuous mire, knee deep to men and horses."[1]

Of course, those profiting from operations on the Dyea and Skagway trails competed to attract customers, and the competition gradually led to improvements in both routes. First one would gain an advantage, then some modification would swing favor back to the other. With the seesaw of conditions, stampeders who picked their route by hearsay actually could have an advantage.

6–1. A section of the Skagway trail that was improved by a bridge in the fall of 1897. (U. W., LaRoche photog., 2070)

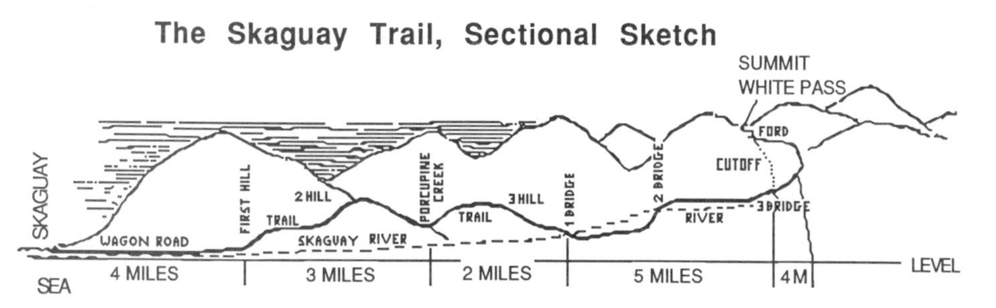

6–2. Cross-sectional map of the Skagway trail as far as White Pass showing elevation changes. Drawn about 1898. (U. W., pamphlet file N979.8)

6–3. Lower section of Skagway River Valley looking north toward White Pass from a spot on the White Pass trail. (U. W., Hegg photog., 354)

By late fall and early winter of 1897–98, bridges as well as sections of corduroy road were in place on White Pass trail, so it regained its popularity. But soon, there was a tramway built to ferry goods across Chilkoot Pass. This speeded transport on Dyea trail because pack animals or heavily loaded sleds could be used on either side up to the tramway. As a result, the Chilkoot route again became more attractive. Furthermore, Skagway's unsavory reputation was no drawing card. Many travelers felt safer going by way of Dyea and avoiding Skagway's gang of con men. But then, construction of a narrow-gauge railroad was begun in Skagway on May 27, 1898. By the end of the summer of 1898, it ran thirteen miles up the Skagway valley, and by February 18, 1899, it was completed to the pass. The train at first gave the advantage back to White Pass trail. But eventually it put both trails out of business.

To reach White Pass summit, eighteen miles from Skagway, was a three-day journey if weather conditions were good. This, of course, was one way. If a stampeder were carrying his or her own goods rather than having them packed, the journey was repeated fifteen to twenty-five times to get a whole outfit moved, as on Chilkoot trail.

The first five miles were on what became a wagon road. The trail

6–4. Tired stampeder on Skagway trail. (U. W., Hegg photog., 553)

forded the Skagway River several times, alternating marshes with very rocky footing.

When the trail left the valley of the Skagway River to cross Porcupine Creek, it climbed a mass of boulders one stampeder likened to "a mountain of goods-boxes, some of them being bigger than the rest—the size of tents."[2] Few had time or the inclination to appreciate the rugged beauty of the area. It just seemed a grueling way to have to carry a ton of goods. The bare rocks of the mountains were treacherous—slippery and smooth, and made more slick by frequent rain, mud, and wet vegetation. Stampeders just wanted to be through them quickly. Tappan Adney came across one exhausted traveler on Porcupine Hill who was "asleep on his pack, with his closed eyes towards the sky and the rain pattering on his face, which was as pale as death. It gave me a start, until I noticed his deep breathing."[3]

After crossing Porcupine Creek, the trail eventually wound its way down to the Skagway River again. Thirteen miles from Skagway and just before White Pass summit, a semipermanent camp sprang up, White Pass City. The camp was at the end of a second day's journey and served as a staging area for the next day's march over the pass. The last few miles to the summit were a steady but not hazardous climb over which pack trains or sleds could readily be used.

After the summit there were seventeen miles of gentle slopes, loose gravel, marshes, and sand before reaching Lake Bennett.

6–5. About 3.5 miles from White Pass Summit just past White Pass City. Caches of goods are alongside the trail, ready to be ferried up to the summit.(U. W., Larson coll'n, Hegg photog., 194)

Transporting up to two thousand pounds of goods across the difficult trail was a major undertaking no matter how it was accomplished. Those stampeders who couldn't hire a pack train either carried everything on their backs or devised other ways to get their outfits to Bennett. One solution, at least for the first five miles out of Skagway, was to make a narrow, two-wheeled cart. This was pulled by a single horse or by a person, with a partner balancing the load from behind. With a cart, stampeders could increase their carrying capacity by some 600 pounds. But of course, once at the incline at Porcupine Hill, there was no place on the narrow, boulder-strewn ledges for a cart.

Tappan Adney observed a surprising test of the durability of one of the little carts while in Skagway in mid-August, 1897.

> One horse, when ready to be loaded beside the scow [in Skagway harbor], became frightened, and started up-town with the cart behind him. He ran into the town, then turned at right angles, crossed a branch of the Skagway, started, cart and all, up the face of the mountain, turned around, recrossed the river, and came back to the scow, the cart now running right side up; then, striking a root and bouncing ten feet into the air, it landed upside-

6–6. Approaching White Pass Summit, seen in the gap to the right of top center. (U. W., Hegg photog., 245)

> down. The cart never ceased for more than a moment to run along, right or wrong side up, on its wheels; not a man was hurt nor a tent-peg torn up, and it all took place in full view, and the crowd greeted with a shout each time the cart flew up and landed all right. A moment later the incident was forgotten.[4]

It was probably not the last time that this poor horse would be frightened. Even though "designed for packing," White Pass trail was a nightmare from the point of view of an animal. The Skagway Valley's natural conditions themselves presented many hazards—little feed, freezing storms, and slippery rocks and mud. Many animals broke a leg when they lost their footing or slipped off the trail and over a cliff.

But the natural hazards were aggravated by humans. Stampeders on White Pass trail were no more familiar with outdoor life than those on any other approach to the Klondike. So problems for the horses and mules were amplified by inexperienced handlers who packed too much onto one animal, distributed the weight unevenly, or failed to feed or care for their livestock properly. Even experienced packers would cut down on food in hopes of carrying more of a pay load. Between the negligence of those who knew better but were concerned more with profit than with the welfare of

6–7. Pack train between the summit and Lake Bennett on White Pass trail, September 6, 1898. According to Edith Larson, this is the American journalist Nelly Bly with her secretary. (U. W., Larson coll'n, Hegg photog., 3193)

6–8. A sturdy two-wheeled cart similar to the one which bounced through Skagway. (U. W., LaRoche photog., 2010)

6–9. Dead pack animals along a particularly hazardous section of the White Pass trail in the fall of 1897. (U. W., A. Curtis photog., 46112)

6–10. Carcasses of dead horses and mules washed into the Skagway River. (U. W., Larson coll'n, Hegg photog., 3101)

their animals, and the ignorance of those who had never handled animals before, several thousand horses, mules, burros, and ponies died. Others were simply shot or turned loose to starve to death at Bennett when their owners were finished with them.

Emma Kelly, who had made it across Chilkoot trail to Bennett by challenging her packers to keep up with her, told of her experience with a pack train arriving on the White Pass trail.

> . . . A most heartless man, who had a pack-train of horses, arrived over that trail, the horses all carrying heavy packs. The animals had not had any food for days, as there was nothing upon which they could graze, and even had there been, their owner was so cruel and avaricious for every dollar to be had, that [it] is doubtful if he would have allowed them time to feed.
>
> They were simply racks of bones, scarcely any flesh upon their ribs or bodies, having made several trips over the Skaguay trail during the summer.
>
> About half a mile from the place where they were to be unloaded, the five in the lead of the train gave out, completely, and one which had fallen down was severely beaten with a rough stick in the hands of the cruel owner. The poor brute staggered to its feet only to take a few steps, when it fell a second time. The other four, as they stood trembling and swaying from weakness and hunger, seemed ready to drop dead.
>
> I watched the man as along as I could stand it, then ran over to my war-sack, got my revolver, and while he was at the rear of the pack-train, shot the animal in the head. I knew it was only a matter of a few hours before the poor beast would succumb, and I thought death would put it out of its misery.
>
> On coming up, the enraged owner started for me, using some very profane language. He was even about to strike me, but I had twenty-two men to defend me and make him swallow every word he said. The only thing I was censured for by my companions was that I did not shoot the man.[5]

Stampeders crossing the White Pass trail late in the fall of 1897 or the following spring and summer would complain of having to step through decaying carcasses mired in the mud. And the stench along the trail was nearly unbearable. In the spring when heavy rains and melting snow flushed the Skagway valley, hundreds of decomposing bodies were washed into the river and floated into town. And so, White Pass trail earned another name—Dead Horse trail.

Flora Shaw (Lady Lugard), colonial editor and correspondent for the London *Times,* crossed White Pass trail in July of 1898. Her baggage and

6–11. A portrait of Flora Shaw (Lugard) after she returned from the Klondike. She was 45 years old when she traveled to Dawson in the summer of 1898 to report on the gold rush. (London *Times*)

provisions were transported by pack train, but she walked the major part of the trail carrying a light pack.[6] In her address six months later to the Royal Colonial Institute in London, she reported that she was the twenty-seven thousandth person to go over the passes, and that her trip was made quite easy by those who had gone before.

> The walking was at times very heavy. If rain had lately fallen it was through pure swamp. Sometimes ankle deep, sometimes knee deep, one was forced to wade along the valley bottoms, the summer sun beating hot upon your head. At times a rocky shoulder of the hill would project itself across the way, and then wading was exchanged for climbing, which had sometimes to be done with hands and knees. Through the valley bottoms streams ran with many windings, and in a country of no bridges when water had to be crossed it must be forded, unless some traveler handy with his axe had passed before you, and the slim and slippery stem of a tree felled and thrown from bank to bank may offer a precarious chance of passing without a bath. Twenty miles of such walking would fill my day from dawn to dark.[7]

So despite extensive improvements, the Skagway trail was still in poor condition in the summer of 1898. But all was not misery on the trail. Here, as on the Chilkoot route, there was often anonymous but genuine goodwill shown by travelers for each other, despite competition to get to the gold as quickly as possible. Flora Shaw was treated with the respect and deference generally accorded women at that time. In addition, her male fellow travelers obviously went out of their way to see that her journey was safe.

> Sometimes in the course of these heavy walks it would happen when men had passed me, talked for a few minutes and gone on, that three of four hours later I would reach some difficult place and find one sitting there resting his pack against the trunk of a tree. "I thought of you," the greeting would be, "when I came to this place, and I thought maybe you'd want a hand over, so I waited for you." One day I chanced to be specially tired, and an extremely rough-looking man overtook me. After some conversation he said, "You're a bit tired; I can see that by your eyes."
>
> "Yes," I said, "I'm tired."
>
> "I expect you're pretty well dead beat."
>
> "Oh, no," I assured him, "I'm not dead beat; I shall get to the end of the day's walk all right."
>
> "Well," he said, "maybe; but I guess I'm going to walk along with you." And he did for twelve miles more, though it delayed

> him several hours, and brought him in late in the evening instead of the middle of the afternoon to camp and food.[8]

Flora met and talked with a number of men when their paths coincided. Whether explicitly through her questions or spontaneously, their reactions to her indicate that she aroused sentimental thoughts of home.

> A large number of the men were married and had wives and children in the outside; and there was a pathos, not easy to express, in the readiness with which well-thumbed photographs would slip from mud-encrusted side pockets, to show to a perfect stranger the shape in which thoughts of home were journeying through the Yukon. Sometimes the picture was of a child, sometimes of a young wife, sometimes, more touchingly, of the middle-aged companion of a lifetime; and I might chance to hear that it was hard on the "old missis" to be left again. All kinds of men from every class of life were there. Americans, Canadians, Australians, and Englishmen were in the majority, but almost every European nationality was represented. One Frenchman, who had lost his entire outfit by the overturning of his boat upon some rapids, and had not even a blanket to lie down in, had saved a curl of his baby daughter's hair. He was cheerfully content, "Ma foi! I have got the thing I valued most!" And more than once the little packet that looked to ordinary eyes like a skein of yellow floss silk was pulled from his trousers pocket for me to see.[9]

Despite her own daring in winning renown as a journalist, Flora Shaw expressed very traditional views about women. Hers was a time when many women were finding a public voice and beginning to see themselves as having a valuable place in history even outside of their reproductive and family roles. And yet, when Flora Shaw married, after returning from the Klondike, she gave up her own very influential career and in arguing against the writing of her biography, she claimed that any woman who had not borne a child was a failure, and her life not worth recording.[10] In Flora's speech to the Royal Colonial Institute, she indirectly acknowledged the commonly held belief that the only females in the Klondike were prostitutes and dance hall girls. Without refuting that idea, she offered her own impressions about what women might accomplish there.

> The question of whether women that men respected could be brought into that country was one of perpetual discussion. Nowhere does one see so plainly as in districts of new settlement the need of woman as a home-maker. The majority of the men in the Klondike, excepting, perhaps, the very young, were in the literal

> sense of the term, "home" sick. They wanted a place as much as a person, but it needed a person to make the place: some one to minister to the common needs of life, to clean the spot in which they lived—even though it were only a tent or shack—to wash the clothes, to cook the food, to give to one's fireside a human interest which should make it, rather than another, the magnet of their daily work. The rougher the man the more imperative the need appeared. The absence of homes in such a place as Dawson explains to a great extent the existence of saloons; and in noting the contrast between the splendid qualities exercised in the effort to acquire gold and the utter folly displayed in the spending of it, it was impossible to avoid the reflection that in the expansion of the Empire, as in other movements, man wins the battle, but woman holds the field.[11]

It is surprising that Flora characterizes women as being "brought into that country" rather than going there under their own initiative, as she herself had done. Many women in the Yukon did fill the roles of home-maker and civilizer that Shaw envisioned. But many women tried other enterprises as well, such as mining, banking, doctoring, storekeeping, restaurant and hotel management, politics, and writing, adding their own unique contributions to each of these professions.

6–12. Stampeders crossing the White Pass trail by sled. (U. W., Goetzman photog., 117)

Ethel Anderson (Becker), the five-year-old who had picked apart a blanket with her brother on their rough voyage up the Inside Passage, had another perspective of White Pass trail when she crossed it in a wagon two months after Flora Shaw. The weather had turned cold by September, and Ethel remembered its consequences well.

> We arrived in Skagway, a peaked mother and three small children. Construction of the White Pass and Yukon Railway had begun. I do not remember much of the trip to Whitehorse, except that I huddled in a tram car wrapped around in an icy wind. A little red hood and wool coat were poor protection. My teeth were chattering when a man lifted me down and showed me how to slap my arms across my body to start the circulation. We danced and jigged and swung our arms until I was warm again, then walked hand in hand behind the slow tram car for some distance.
>
> Somehow mother fed us and washed us, diapered Clay, and soothed him to sleep with a sugar tit, but it was an experience she would never talk about in later years. Hundreds of men turned back on that White Pass Trail, not even a fortune in pure gold being worth the hardship. But that pioneer mama of ours was re-uniting her family and building a new home.[12]

The Feeros, who were prospering in Skagway, moved up the trail to White Pass City during the summer, where they had another little cabin. In addition to the pack train, they also ran the White Pass Hotel. The hotel consisted of a restaurant and a couple of bunkhouses where travelers could get out of the weather for a while. Emma Feero's cousin, Annie, cooked for the White Pass Hotel and also helped to take care of the family.

Summer days for the Feero children at White Pass City were spent roaming the camp and the surrounding hills while the stampede swept by. In the camp they could find adults who would oblige them with stories while they waited for supplies to be loaded at Lilly's or Bledsoe's feed stores. Or they would steal pies when Annie wasn't looking and hide themselves in a bunk house while they hungrily devoured them.[13] And sometimes they would recognize among the stampeders someone they had known in Skagway. For example, there was a woman who had been in Skagway in the winter of 1897–98, who reappeared, transformed, in White Pass City the following spring.

> . . . A kitten had come to us, and we took it in. We thought we had a prize. We had a dog, and we had a few chickens, and if we had a cat, we had a whole farm! Well, [an old lady] came along the street one day and saw the kitten, and she grabbed it away, and she says, "You got my cat." Took it away!

6–13. White Pass City. The Feero cabin is at bottom right, and the French cabin is on the hillside at left where the wash is hanging. (Edith F. Larson coll'n)

We watched her all [the] time then, see when she was going to leave, because she wouldn't take that cat with her over the trail. She was in the bend of the road where we had to go right past her tent. And we'd go past the tent, and a little piece of her skirt would be a little shorter. Told Dad, "Now we'll get the kitten because she's got her skirt shorter. Goin' over the trail." When she got the skirt up here to her knee, then we knew we'd get that cat. Then Father said, "Don't . . . count too quick!" When we had passed the place, the tent was gone, the cat was gone, everything was gone!

. . . Nobody ever found out what she done with the cat. But she got to White Pass City and stopped at this hotel. The man father had workin' taking care of the men's part came in one night and said, "Annie," that was mother's cousin, . . . the cook, "I got something out here, and I don't know what to do with it. It looks like a man, but it acts like a woman. I don't know what to do with it." So Annie says, "Bring it in the kitchen, and I'll find out what she is." She got in the kitchen, she was soakin' wet, and she spread her knees out. "Oh, my pants are wet!" Annie says, "Tommy, go on, I'll take care of her." It was a woman. This was the same woman as had the cat.

> . . . She put on men's clothes and acted like a man because she thought she'd be safer. Right there in the first thirteen miles when Annie discovered she was a woman, the stampeders found out she was a woman. They helped her all the way through to Dawson. And when Father went into Dawson in September, she was peddling apples from saloon to saloon . . . and making a good living.[14]

Women were not always welcomed by those running roadhouses on the trail, as Florence Hartshorn discovered. In July of 1898, she was on her way to join her husband at Log Cabin, a small settlement toward the end of White Pass trail. (Near that location today, the Skagway-Whitehorse road crosses the White Pass and Yukon Railroad line.) When Florence stopped at the tent roadhouse at White Pass summit on her second night on the trail, she got an unexpectedly cool reception.

> On reaching the place where a door should be, I pulled back the flap of the tent and caught my first glimpse of the lady who had been recommended as being a very nice lady. I stood waiting for her to turn. At last I cleared my throat and she turned and gave me an icy stare. I said, "How do you do," but she didn't reply for a minute. Then, "Well, what do you want?" she asked.
>
> I asked her if she had any accommodations for ladies and she very tartly said, "No, we have no accommodations for—women." I immediately caught the inference on "Women of the trail." . . .[15]

Florence attributes her lack of welcome to women who were mining camp followers—prostitutes or "sporting women." According to her, the sporting women were a source of much trouble on the trail, especially for roadhouse proprietors, and this one at White Pass had decided that none would be permitted in her house again. Still, Florence was respectable and needed a place to stay the night, so she persisted.

> "Where else can I spend the night?" I asked and she said with a shrug, "Over at the Police Barracks." I told her I hardly wanted to go there and asked if I might not sit in the chair. "I should say not," was the reply. I sat there for a time anyhow wondering how else I might approach her.
>
> She took down a small can from the shelf and took out a spoonful of desiccated eggs, placing it in a cup. Then she added the same amount of water and set it aside. A cup of sugar was placed

in a stirring dish together with a dash of salt. My woman's curiousity aroused, I asked her if I might not watch her as soon I would be doing the same thing for my husband was opening a restaurant at Log Cabin.

"Really," I said, "I haven't any idea what it will be like." Seeing that she was listening I told her that my husband was the blacksmith at Log Cabin and I was supposed to have been at Log Cabin that night, but that trouble had hindered us.

Just then her son came in and she asked him if he knew the blacksmith of Log Cabin and he said, "Who, Bert?" Then seeing me, "I bet you are Bert's wife. Bert was worried when you didn't come."

Then the lady stopped making her sourdough bread and grabbed me and hugged me. "Only this morning I vowed that not another woman should spend the night here! Son, you tack blankets around the bunk, rob all the other bunks of their pillows and make her a feathery mattress." The lady was very nice to me and I had a good night's sleep.[16]

Not everybody who made it as far as White Pass City continued across the trail to Dawson. Edith made the acquaintance of one family, the Frenches, who eventually simply returned to Seattle.

6–14. The French camp on the Skagway trail in 1897. These may be the Frenches to whom Edith Larson refers. (U. W., A. Curtis photog., 46028)

> You see the [cabin] up there with the clothes hanging down on the side? [See Figure 6–13 of White Pass City.] That was two brothers from Seattle, and one of them had his wife. . . . You know what nosey kids are. We was up on the hill playin . . . , and they were up there cutting wood. So we caught up with them and said, "Why don't you go on to Dawson like the rest of the people do?" Well, the good Lord came to them and told them not to go any further until He came back. So they stayed there. Till they ate their grubsteak up and all the rest of the . . . stuff they could find, and then about 1900, they went back to Seattle.[17]

Packing on White Pass Trail

Because her father operated a pack train on White Pass trail, Edith Feero (Larson) was most familiar with the events on that route and with operations of professional packing. Pack trains left regularly from the Feero cabin on the north end of Skagway, so from Edith's perspective as a child, the trail began and ended at her doorstep. John Feero worked along the trail only a short time—from late July of 1897 to December of 1898. But this was the period when the Skagway River Valley was most changed by the gold rush.

To run a pack train, you had to have animals. According to Edith, many of those used on White Pass trail were wild horses brought in from the States.

> Those horses were just picked up off the prairie down [there] in the states, and drove onto a boat and hauled up here and turned loose. That's where our horses came from. Mules, most of the mules came from California. They were brought in here by a family by the name of Dewitt. And they had, I think he had 100 mules. . . . They'd just bring 'em in there and unload 'em in the bay. They'd swim ashore. And then they'd just haul 'em off. You'd buy so many horses, then they drove 'em up to your place.[18]

Of course, getting a wild horse to be an effective pack animal was a bit of a trick. This is when the experience of skilled horse handlers paid off.

> They bring in these wild horses off the boat and run them up and lasso 'em. Hang 'em by the head [next to a tree], the stump's cut way up high, and tie their heads way up high. Blindfold 'em, and pack 'em. If you got sugar, eggs, flour, anything like that on the wild horses, they don't stand and take it! They stand and look around, and look and see if there's a place they can go through—that they themself [and nothing else] could go through. And they

> go straight for it! Well, all your perishables is gone. So they learned the lesson early. Pack those wild horses with hay. If they break it up, you feed it to them. Don't put grain on 'em. Don't put any perishables on 'em.[19]
>
> ... One morning we got up out of bed, got up to go to school, and the whole yard was full of wild horses! They had 'em all tied. We had one man in the outfit that could break anything, but he never hurt 'em. He could take a sore-backed horse and pack it and cure it. He could do anything with a horse. But he was the only one that you could trust to do anything. He'd take them horses, and he'd tie 'em up and pack 'em. There was one little animal that bucked every single morning. Never failed.... So they put the cooking utensils, fastened them together good, tie 'em on top of the pack. So he starts to buck, it will rattle. And he has to get his rattle all out![20]

Whether a pack train made it to Lake Bennett and back without loss depended a lot on how the animals were fed. There was little if any natural feed along the trail. In the summer there were occasional meadows in the Skagway valley, but these were soon grazed out with the earliest stampeders. In any case, grass alone is not sufficient to nourish a hard-working trail animal. And yet, whatever food was packed in displaced payload. So many tried to minimize what they fed the animals on the trail. They figured it was cheaper to use a horse until it dropped, then buy a new one, rather than to buy and pack in feed. This callous treatment of the animals was one point that John Feero just couldn't accept. As an experienced horseman, he knew what it would take to run a successful operation. And perhaps the fact that his family of observant children were watching the realities of the stampede made him all the more determined to ensure that his own involvement was under humane conditions.

> My father took the first pack train over the trail that never lost a horse. He [was working for someone else at first.] ... They got 7 miles out on the trail, and they fed their horses a double handful of oats, like that [cups two hands together]. Seven [miles] on the trail and a double handful of oats. They had their own dinner, went to bed, they got up.... Morning and they give their horses another double handful of oats. Started packing. [When they got a little further], my dad said, ... "Gonna feed your horses?" [The foreman said,] "We already fed the horses!" "I'll never work for a son-of-a-bitch in my life that would starve his horses. And I'm not workin' for you!" The foreman says, "You got to go, because you signed up." ... [John] said, "I'm goin' this way," and he went back into town.

[The owner] asked him, "What are you doin' in town, John?" He says, "I never worked for a son-of-a-bitch in my life that'd starve his horses, and I ain't working' for you!" "Well," [says the owner], "did you ever cook?" My dad says, "Yeh. I used to cook when I went with my brothers on the river drive back in the east." "All right, cook dinner. The train will be in." He cooked the dinner, and they were sitting at the table, and [the owner] walked past there and . . . said to the foreman, "You're fired! John's going to take the train out next time."

Well, he had never told Dad he was going to put him on [as] the boss. . . . Next morning they got up and loaded the horses. He says, "John, this is your train." "Where's the feed?" "Feed's on the horses." . . . "I want three horses, and I want a man, extra. I want them three horses packed with feed, and that extry man when the feed is eaten up, he brings the horses back, those three." "We can't afford that! We get fifty cents a pound. . . ." He says, "All right, I ain't goin'." "Well," [the owner later reported], "I had a long time gettin' him to go, but before he went he had the horses and he had his feed."

There was a pack train out on the trail then four days. [John] went and took [his] outfit and went to Bennett and came back with *every* horse. The other pack train came in two days later, and there was twelve horses to start off with, there was four came back.

So you see, it was feed. Most of them horses starved to death. Some broke their legs, and then they have to kill 'em, 'cause you can't [save them.] Then some of them . . . committed suicide. [Father] told Mother about animals committing suicide. He said, "I can't believe it!" But one night he came home and he says, "It's true. I saw it. That horse went along the canyon, rubbin' its head on the trees as it was goin' along, till it caught its . . . halter rope in the tree. Just stepped over the bank and hung itself! Now I know that's true now."[21]

Getting warm, nourishing food on the winter trail was a problem for humans as well as for the animals. Fortunately, keeping food preserved wasn't hard in the mountains' natural deep freeze.

See they always took beans and freeze 'em and put 'em in a flour sack. Mother took up a lot of corned beef when she went up 'cause we had a little steer. She had it killed and corned. And cook a lot of corned beef, cook rice, put raisins with it. Things like that. Then when it come night, they'd take and chop off as much as they wanted for the meal. Keep the rest all frozen. And

> that was all tied on the top of the horses, away from the heat, on top. And that's the way they kept their food.[22]

John Feero had started his own pack train by October of 1897. Some of his first horses were mistreated castoffs of other trains. Besides feeding his animals adequately, Feero also improved his methods by using a sturdy lead pony to guide the rest of the animals.

> His first little horse he bought was a little black mare. And she was little! We kids at ten years old could look over her back. He bought her because she come in and fell on the ground. They had lost 8 horses out of a pack train of 12 that went out, and this little horse was one of them that came back. And he said, "I'll give you five dollars for that horse." "Take it! It's going to die anyhow." She was the best horse in the train that winter. Kindness took care of it.[23]

The little horse was named Nellie, and she proved her worth many times over. Ponies or small horses were more sure-footed than larger horses, so were better at picking their way through the hazards of the trail. Tappan Adney, the *Harper's Weekly* correspondent, was surprised at the difference he saw in how ponies handled the steep, slippery rocks compared with horses.

> . . . The lank, big, clumsy horse is in danger at every step. He comes to a drop-off, lifts his head in air, tosses his forefeet ahead with a groan, and trusts to chance to find a footing. He strikes a sloping rock, flounders for a foothold, and down he goes sideways and rolls over. . . . The little cayuse, or Indian pony, however, like the mules, looks where every foot is placed. One cayuse got out of the train and came to a pitch-off of ten or twelve feet; we looked to see it break its neck, but it simply put its head down, slid over the face of the bowlder, and landed squarely and lightly as a goat.[24]

In the winter of 1897–98, there was so much snow that the trail was blocked and all commercial packing was stopped for a time. When the trail could be opened again, it was little Nellie who led the way. This same trip also gave one of the prostitutes in Skagway her chance for a firsthand look at trail life.

> Well, the packers downtown would go, but they needed a leader. Little Nellie, the one that my father bought for $5 was the best leader on the trail. Then he had Maggie—that was another fairly good leader. [One packer] said, "If you can get John to go and take Nellie, we'll go, but we won't go unless we can get Nellie to go."

6–15. Switchback on a section of the treacherous Porcupine Hill where many pack animals lost their lives. One pack horse is barely discernible here amidst the rubble, in the upper center of the photo. (U. W., LaRoche photog., p. 64, 2067)

Well they got it all figured out so they was going to go, and then this sporting woman, they called her the Skagway Belle, she wanted to go when they broke the trail. So the night before they were supposed to go, she come up to the cabin and knocked on the door. Afterwards, Dad come and he said, "You know who that was?" Mother said, "No." He said, "That's Skagway Belle. She wants to go with us. But all the packers told her she couldn't go unless she gets permission from John Feero. . . ."

So she says, "You know what I am. I can be a lady on the trail. I can help warm up your meals, and I can help you what I can." So he says, "Well, then we'll let you go. But you're a lady on the trail."

They got up to the foot of the summit, just past . . . White Pass City, and the storm come down on them. They couldn't unload the horses cause the stuff would get all covered up, and they couldn't let the horses lay down. That Skagway Belle went around with the men all among them horses, they had a lot of horses up there, pattin' them, keeping them on their feet. Didn't dare let 'em lay down. About 10 o'clock at night the storm cleared. . . . It was so darn cold, it was a blue cold. And she walked among them horses all that night until it stopped. And then the moon came out and then the stars. It was the most beautiful thing that anybody could ever picture! That snow was perfectly clear, flat, smooth as anything. And all these horses, one by one, they put 'em on the trail and they went up.

They didn't have no pack on Nellie, and no pack on Maggie, because Dad wouldn't let 'em put a pack on Nellie. But they had her bridle on, they had her saddle on, and when that poor little devil came home, she had hardly any hairs on her tail and her mane. When she'd slip off of the trail—it was slippery so she'd slip off—they'd get her by the tail and the bridle and the saddle. She was little, so they'd pull her back up on the trail. Set her down on the trail, and she'd go on until she'd slide off again. That's where she lost her hairs.

But that was the true story of Skagway Belle. And she went over the trail with them and came back and was a lady all the way.[25]

Besides the outfits of stampeders, many of the supplies for Dawson were packed in by the White Pass trail, and much of the Klondike's gold came out the same way. So there was often a surprising assortment of goods moving through White Pass City, from whole canoes to live turkeys and chickens, as Florence Hartshorn at Log Cabin described.

One day I heard a strange noise outside my cabin door. I went to see what the strange noise might be and I could hardly believe my eyes. I had seen strange sights but this was the strangest of all. For I saw turkeys on all sides. They were running everywhere, nearly 100 in all. A few men were herding them. These birds had walked over that White Pass Trail where men had met all kinds of hardships but the turkeys had experienced no such hardships. No trouble at all. The hills were covered with berries and they had all they could eat. At night they were ready to roost away up in the tree tops. It was strange to see how the men kept them together but they said they had no trouble. They had already made over 30 miles—had eight miles yet to go before reaching Bennett. Here they would be loaded in scows and floated down the Yukon River.

The men slept outside to be ready to start when the turkeys decided to come out of the trees for breakfast which was very

6–16. Pack train with crates of turkeys along White Pass trail, 1898. (U. W., Larson coll'n, 9087)

early. The nights in summer are light so the turkeys had no way of knowing when it was time to start. The men said the morning was the hardest time as the turkeys would separate but when they were on their way they could keep them together. I asked how much they expected to get in Dawson and they said, "$1.50 per pound and people are waiting for them."

Another time I saw 1,500 turkeys that crossed the trail in crates on the backs of horses. Their heads stuck out between the slats as they passed along. They could not enjoy the nice red berries as those who had walked had done.[26]

Edith Feero Larson also told of some unusual shipments on the trail.

. . . One day father took a contract to take twenty tons of bonded whiskey over the road. That was in June of '98. And that twenty tons of bonded whiskey went through our yard from noon of one day to the noon of the next day. It was on horseback or on wagon. The wagon was open then. And that night my brother—I think it was on the twentieth because . . . it would be daylight all night, you know—my youngest brother wanted to see it get dark. And he set up, and he was going to see it get dark. About 1 o'clock in the morning he opened the door and he says, "Daddy, it ain't gettin' dark, it's gettin' light." Dad says, "You're sleepy. Come on in and go to bed."

> But that night . . . the pack train next to us lost a horse and driver. . . . They couldn't find that darn horse and they couldn't find the driver. The driver finally woke up and he came in. But he couldn't find the horse! He took the horse and hid it down along the river bed so he could steal some of the whiskey. And off he went to sleep, then he didn't know where the horse was. He never got the whiskey. And they didn't find the horse until along in the afternoon of the next day.
>
> . . . One time they come in there with $30,000 worth of gold. Packed it in . . . from Dawson. And Mother run out—she was mad. They had tied those horses together, what they called "tailin' them." You tie one horse to the tail of the other, see. Mother says, "John has told you time and again not to tie them horses. They got to be free."
>
> "Well," he said, "the little devils wouldn't stay together." And he take one little pack off and he said, "Catch this!" And he throwed it to her. Well, she couldn't catch it. It was gold. It was heavy. Fell on the ground. That gold, they had seven horses with these, and they were all heavy with gold. . . . This was the mail train. They throwed that all in underneath the window in the front cabin. You see there [were] two cabins there. Throwed it in there under the window. And left it! Nobody to watch it. Nobody stole anything.[27]

There was a code of honor on the trail, and most stampeders followed it. But there were scoundrels too, most of whom seemed to gravitate to Skagway. One explanation often given is that the Northwest Mounted Police were simply stricter and sentences tougher in Canada, so crooks soon learned to work in the United States. Or maybe it was just that Skagway was easier to reach for crooks as well as for stampeders.

Skagway's reputation for lawlessness is epitomized in the figure of Jefferson "Soapy" Smith. His nickname came from a con game in which he, in plain sight, put a five-dollar bill around a cake of soap, then rewrapped the bar and sold it for fifty cents. When the purchaser opened the soap he found . . . soap! Smith's quick fingers had deftly removed the money during the rewrapping process. Soapy had become the unofficial leader of the gamblers and con men who operated out of Skagway, and many stories are told attributing much of the crime in Skagway and on the trails to him or his associates.

> . . . You know they'll tell you about Soapy Smith did this, and Soapy Smith did that. You know what Soapy Smith was? He was a gambler.
>
> You're a gambler, and I'm a citizen see. And I show you I've got

6–17. Con men working over stampeders at shell game tables, probably near Skagway. (Y. A., Vogee Coll'n, 118)

a lot of money. Now it's your business to get that. That's what his business [was.] And I don't think he committed too many crimes.

He had a gambling table about a block above us, and passed back and forth by our place all the time. He had another place downtown. So this one up here, they had what you called "boosters." And they'd come down here. Maybe you're getting ready to go out. "Oh, what are you doing? Where are you going?" They're all packed up with a great big pack on their back, all doubled up like they have a ton on their backs. Well, you got to get out and get a going. "See what I got! [showing some money] I got this off this little table up here." The sucker goes up to that table, and leaves his money. See this is the booster. That big heavy pack he's got is nothin' but light hay.

He got a lot of [people] that way. One day they come in and they said, "John, don't go out today." "Why?" "Well, Soapy Smith has got the road blocked up the canyon. You can't go through." He said, "I'm going to go on." "You can't get further than the canyon." "All right," he says, "I'm going to go anyhow. They won't bother me." Packed up and took his load and went on past. If you didn't stop, they didn't stop you. You either had to be a sucker, or you didn't get caught.[28]

6–18. The Fair View Hotel in Dawson. Belinda Mulrooney's masterpiece was located at Front and Princess streets and overlooked the Yukon River. (A. H. L., Wickersham coll'n, Larss and Duclos photog., PCA 277–1–115)

Soapy Smith was a complex character. Though he was the acknowledged leader of the con men in Skagway, he often carried out many fair-minded deeds that won him the gratitude of those he helped. One of those was Belinda Mulrooney, the enterprising woman who was among the first to cross Chilkoot Pass in the spring of 1897. In one year's time she had made enough money on restaurants, roadhouses, and mining to build Dawson's finest hotel, the Fair View. As she explained in an interview many years later, she was ready to furnish her hotel in the spring of 1898.

> The only way to get [the shipment of furnishings] was to go out and bring it in myself over the White Pass. I ordered it all, cut-glass chandeliers and silverware, china and linen and brass bedsteads. I signed a contract with a fellow named Joe Brooks, a thick-set Vancouver drayman who had some mules and a pack train, and paid him $4,000 down and the balance at Lake Bennett to pack my stuff in.
>
> Brooks took my freight out a mile along the trail and dumped it there for a better offer to haul whisky. [He also did not return the down payment.] There was a clause in the contract that if he didn't deliver the pack train was mine. He just laughed at me and

6–19. Joe Brooks, pack train operator on White Pass trail, on his pinto horse in about 1898. (U. W., Larson coll'n, 9080)

said, 'What are you going to do about it?' He was fond of drink and had so much money he didn't know what to do with it.

I went back to Skagway and I was boiling mad. Soapy Smith was clean and he was intelligent-looking. I thought he was the most perfect gentleman there. He looked like a minister and had this soft Southern drawl. Soapy was my hero, and I went to him for advice. I told him what had happened and how I owed $8,000 and had to get this gear into Dawson.

"I'll see what I can do for you, Belinda," he said.

He picked up a good tough bunch of men and we lit out with his crew and took possession of the pack train, unloaded the whisky and packed my freight. Brooks had a pinto he rode all the time and I took that for myself. That's what Soapy did for me and I liked him.[29]

The White Pass and Yukon Railroad

While the stampede proceeded on foot and hoof through the spring and summer of 1898, construction of a narrow-gauge railroad was also under

6–20. Work on railroad bed about six miles from Skagway in early summer of 1898. (U. W., Larson coll'n, Hegg photog., 9081)

6–21. Laying tracks down Broadway in Skagway, June 15, 1898, between Fifth and Sixth Avenues. This view is looking south, with the beach at the end of Broadway to the right. (U. W., Larson coll'n, 9084)

6–22. First train in Skagway, July 20, 1898, sits on newly laid tracks. (U. W., Larson coll'n, 9086)

way. Soon the railroad would make the trails to Lake Bennett obsolete. Edith Feero (Larson) described the progress.

> Now you see how fast the country grew. First horses landed there in July of '97. . . . From the trail . . . started in August of '97, . . . they moved down onto the river [in the winter.] In the spring they moved up on the wagon road. In the fall of that year [1898] the train was going up to White Pass City. . . . We spent the summer in White Pass City. And my mother brought my brother [who was sick] down from White Pass on the train.[30]

The railroad was an international undertaking. Its route, of course, crossed territories of both the United States and Great Britain. (Canada was a colony of Great Britain at the time.) Financing was provided by a group of British investors; construction was supervised by an Irish-American contractor, Michael J. Heney; and work was carried out by whoever had the strength and determination to hew the roadbed using only blasting powder and picks and shovels. Edith explained how the railroad, which was to become the lifeblood of Skagway even long after the gold rush was over, came about.

> In February [an English syndicate] had a meeting to see if it was feasible for a [rail]road to go over there. And they said no, no road. There was a man by the name of M. J. Heney, he had worked on railroads out here [in the mainland United States.] He said, "I can

6–23. Cutting the roadbed on Tunnel Mountain near White Pass. (U. W., Barley photog., 222, 9085)

6–24. First passenger train arrives at summit of White Pass on February 20, 1899. (U. W., Hegg Photog., 658)

6–25. Before the railroad was completed all the way to Lake Bennett, goods were transferred at the summit onto wagons or sleds to be hauled the rest of the way. (U. W., Hegg photog., 292)

6–26. A big celebration at Bennett on July 6, 1899, marked the completion of the railroad to that point. Afterward, guests rode on the first "passenger" train from Bennett. (U. W., Larson coll'n, Hegg photog., 366)

> do it." They gave him the chance to try. Landed the surveyors in March. In June, I can't remember the date, they landed the first engine. In August, they were running the train over that first 15 miles, past White Pass City.
>
> And that railroad is claimed to be the only railroad ever built that paid for itself mile for mile as it was being built. . . . Because each time, you could get your freight hauled so many miles, you saved on the packing. That's how they paid them. . . . If you can get your freight on the railroad, then you can go up there much faster then you can with a pack train.[31]

Work continued on the railroad throughout the winter of 1898–99. Tunnels were built and roadbeds blasted out of solid rock walls, all in the midst of freezing temperatures and heavy snowfall. By February 1899, passengers boarded a train that would take them all the way to the summit of White Pass.

There, progress on the railroad stopped. The Canadian authorities in Ottawa informed the Northwest Mounted Police that the railroad should not be allowed to proceed across the boundary that was then established at the pass. In the meantime, the transfer of goods to Bennett did not stop. Professional packers simply started farther up the trail, off-loaded goods from the train, and packed them into Bennett.

The official blockade at the summit was eventually circumvented, so construction of the next twenty miles to Bennett went quickly. On July 6, 1899, the final spike was driven in a ceremony at Bennett. The first "passenger" train then left from Bennett to Skagway, its passengers seated on ties laid across open flatcars. It also carried a reported $200,000 in gold dust from Dawson.

Eventually, the tracks from Bennett were joined at Carcross, at the north end of Lake Bennett, with another set started from Whitehorse. So by July 29, 1900, one could ride the railroad from Skagway to Whitehorse and there transfer to a riverboat to Dawson. No more struggling across the mountain trails. No more building of boats at the shores of Lakes Lindeman and Bennett. No more fighting uncertain winds on the upper lakes. And no more adventuresome shooting of Miles Canyon and Whitehorse Rapids. The route to the Klondike was tamed. Meanwhile, much had happened with the Feero family as well.

Death on the Trail

Packing on White Pass trail was still a very profitable business in 1898. As fall approached, John Feero decided to go to Dawson, as he had planned when he first left Tacoma more than a year earlier. But this time the trip was not to prospect for gold. By the fall of '98, all the creek claims had been staked long ago, and many people in Dawson were without jobs and wondering how they were going to make it through the winter. No, John Feero knew that his gold rush was in Skagway, hauling goods for those who were already in Dawson or who wanted to be getting there. His trip to Dawson in 1898 was to collect an unpaid debt.

> So Father went to Dawson in September to follow [a man who owed him money.] My father packed his . . . chickens in one time for him, packed them over the trail. And that was quite a sight! All them heads comin' out of the boxes! [Later, another man coming back from Dawson told me,] "I come in from the [cricks] one time, . . . and I thought, 'Well, I'll get in and have me a good meal.' Going down the street I saw a chicken in a crate. It was a little bit of a thing, and I thought, 'Well now, . . .' Seventy-five was marked on the crate, see. 'Well, seventy-five cents, that will make me a good meal.' " He went in to buy the chicken. The seventy-five was dollars! He said, "I didn't get the dinner."
>
> [The man my father followed] went out and he got a lot of potatoes and brought them up. My father packed them out to Ben-

> nett. While Father was in Skagway, that man put the potatoes on [a] scow and away he went. Didn't pay [my father.] Father had to go to Dawson to try to collect. Well, he never did get it all. But [the man] said he would get a dollar a piece for them potatoes when he got to Dawson. . . . Some of 'em I understood he didn't sell until spring, but he got rid of all the potatoes.[32]

According to the report of the *Daily Alaskan,*[33] one of Skagway's newspapers, John returned from Dawson just as the Yukon River was freezing. He then proceeded to get his business in order for the winter packing to Atlin, just east of Skagway, where another gold discovery had been made. However, Edith remembered that the preparations were, in fact, so that the family could go out for the winter. And now that they had some money, John could plan to have a much-needed operation. "He went to Bennett to finish his business. We was to go out to the States that winter, and then he was going to have an operation for his hernia. And when he was coming back from Bennett was when he got caught in the snowstorm."[34]

What exactly happened, like accounts of most important life events, is not completely clear. What seems to be consistent among the reports is that the weather was bad, that John became disoriented by the weather and by his pain, and that he died of exposure before help could reach him. The report in the *Daily Alaskan* was as follows:

> [C. W.] Amery and Feero left Skagway with a pack train a week ago and made two trips from the summit to Bennett. On the second return trip yesterday . . . O'Brien's pack train was met at [the foot of the summit] and engaged Amery in conversation while Feero went on [on Amery's horse.] When Amery attempted to overtake his companion he found he had led the horse down a blind gully and was no where to be found. Feero was finally located on a distant hill and upon reaching him the wanderer insisted he was on the right trail and could not be induced to move. Amery then proceeded to pack him to within a half mile of camp eleven and clearing a space in the snow told Feero to continue walking up and down until he went for assistance. When Amery reached Feero, followed by five or six of the railroad laborers he was found to be dead and his body was taken to the camp. . . . Sometime during the year he sustained a severe rupture and this is supposed to be the cause of his death as he frequently complained while on this trip of terrible pain that became so acute he was out of his mind most of the time after Amery found him.[35]

Edith remembered more details about the tragedy.

6–27. The Feero women joined Ladies of the Maccabees, one of the first women's beneficiary societies, formed in 1885. It offered its members insurance for disability, old age, sickness, accidents, survivors, and funeral expenses. Funds came from monthly assessments of the membership.[40] This photo was taken about 1904 and shows Ethel and Edith forming the first row, with their mother, Emma, at right in the second row. (U. W., Larson coll'n)

> I got a picture of that that I can't get rid of. . . . His rupture came out, and he couldn't get it back. That's what caused his death. They was out there in this snow storm. They just padded, he and another fella, padded way back and forth about six or seven feet so they could keep movin' in the storm. And when it cleared up and they could see a smoke, why this other fella . . . said, "I'll go for help, and don't you lay down." He was exhausted himself, so he couldn't keep a straight road. And he got to the hospital—it was the hospital smoke they could see. . . . It was only about a mile away. Told them that there was a rich Klondiker—he was crazy; didn't know what he was sayin'—back on the trail. And to catch the rich Klondiker, they had to follow his tracks. [It took longer to follow the tracks than to go directly, so that] by the time they got there, my father was dead. He wasn't cold yet, but he was dead.[36]

Even 80 years later, Edith's memory of how she and her family were told of John's death was as clear as a photograph.

> . . . Mother and us children were in the cabin. And this man walked in, and he just walked right up in front of my mother and he said, "John is dead." That's all the chance we had. You know how that would take you. And that's the way we got the notice that Father was dead. He just walked in and stood in front of Mother and said, "John is dead." And you know, I can see her, when I stop and think of it, I can see her in her rockin' chair, and that man come in.[37]

Emma was suddenly a widow with four children. And the Feero family, with the prospect of more comfortable life in sight, was thrown again into hard times. "I heard later that my father did have a lot of money on his body. . . . He had sold the horses to the railroad company, and the railroad company had all the horses at Bennett. That's what he went over for. But we never got nothin' out of it. Never got all of his [money owed him from Dawson] either."[38]

So Emma and the children were left to their own resources. With the concerned help of their neighbors, they made a new life for themselves in Skagway. "Well we grew up [fast.] You couldn't be left in a better place, because here in the states there was no welfare, there was nothing. If you didn't have it, you didn't have it. Up there, if you didn't have it, somebody else did, they'd help you. . . . Everybody was the same. And it was always good. And you know, that spirit, in a way, lives through Alaska today."[39]

7

SHOOTING THE YUKON HEADWATERS' RAPIDS

"I don't know when I enjoyed anything so much in my life"

On the south shore of Lake Bennett, where the Skagway and Dyea trails converged, thousands of stampeders paused to build boats that would carry them on their three-week journey to Dawson. Lake Bennett, nestled between ridges of the gray St. Elias mountains, is long and narrow, stretching north for twenty-six miles. It is the first in a series of navigable lakes connected by rivers which compose the upper reaches of the Yukon River. Last in the series is Lake LaBerge (sometimes spelled "LeBarge"), about 160 miles from the head of Bennett. In between, on the rivers are formidable rapids, and on the lakes are shallows, uncertain winds, and hoards of mosquitoes. This path to Dawson is not an easy one.

To build a boat at Bennett in the late summer of 1897, stampeders had to find trees big enough for lumber, cut them, trim off all the branches, haul the logs to the lake shore, then laboriously saw each into boards—and all this by hand. Although there was a sawmill at Bennett, whipsawing logs by hand remained a major source of lumber for those with limited funds. Since even the best timber around the lake was not very large, it took lots of cutting and hauling to get enough material for a boat. To save time and effort, many decided simply to build log rafts instead. And with every boat and raft built and every campfire lit, the fragile land around Bennett was stripped of more of its precious timber.

Winter arrived late in 1897, so stampeders who got to Bennett in the fall worked furiously in the hope of getting to Dawson before the river froze.

On October 1, 1897, the usually high-spirited Mae Meadows who had survived the Sheep Camp flood, wrote a bit more sedately to her sister in Santa Cruz about the tension and activity around Lake Bennett.

7–1. Stampeders whipsaw logs near Lake Bennett. The back-breaking labor of hoisting the saw before each downward cutting stroke and the skill required to keep a straight cutting line led to many strained muscles, short tempers, and over-taxed partnerships. Just behind the sawers are the beginnings of a boat. (A. H. L., Metcalf photog., PCA 34–54)

> The wind blew hard yesterday and snowed six inches last night. The big tent is up and the big heater burning, so I am very comfortable now after our hard trip. There is not much time to rest or any thing. Our last horse died last night, but only have 1,500 lbs to bring from Deep Lake, two and one-half miles, all down hill. Do not know how long it will take to build the two boats. May get a chance to write again. . . . There is a party going back and will bring letters back for 25 cents. It is getting pretty late and it will take a lot of hurrying to get over the lakes before it freezes. . . . The latest report from Dawson is that provisions are very scarce. We have plenty of everything for fear it is true.
>
> Well, I must get out and see the town. I can hear hammers everywhere, it is the busiest little town I ever saw. People are hurrying off every day.[1]

Mae and Charley Meadows did leave Bennett by boat October 18, 1897. They got to within eighty miles of Dawson before they were frozen out of the river, whereupon the indefatigable couple built sleighs and pulled a portion of their goods on into town.[2]

THOSE who crossed the passes in the winter of 1897–98 after the freeze could choose to try to make the rest of the journey to Dawson completely by sled. This, however, was not so easy as it might sound. The Yukon River did not freeze smoothly, as one early resident, Mary Ellen Galvin, explained in a letter home: "When it does freeze up, instead of freezing smooth the huge cakes of ice seem to be standing on edge from 12 to 15 feet high in places. I don't know how to describe it any better than by likening it to an ice-house blown up with dynamite."[3] So the winter trail was overland and rough, and with temperatures dipping below fifty degrees Fahrenheit, it could be extremely hazardous. Nevertheless, some people attempted it. One was the hardy woman described with admiration in the *Klondike Nugget,* who made the entire trip to Dawson on her own in the spring of 1898.

> There is in this city one of the grittiest little women in the whole world. She had heard the stories of the Klondike's richness. Away back in her cozy home on a Connecticutt farm she planned how she would come to Dawson alone and over the ice and thus be the first woman to thus make the trip. She had warm friends in New York, and to them she went and unfoulded her plans. They tried to dissuade her, but to no avail. She proceeded to Puget Sound and there studied out what she would need for the journey, made her clothing and trained her two dogs. She landed at Dyea with 900 [pounds] of freight and proceeded to dog sled it up the trail, and in the face of storm and extreme cold landed it all at the scales. Then from the Summit down over the long stretches of lakes she mushed her dogs on and put her entire outfit to the foot of Lake LeBarge. There she left a part of her outfit to come on by boat and proceeded on, reaching Fort Selkirk when the ice was getting too soft and dangerous for further travel. She pitched her own camp, cut her own wood, cooked her own vituals, loaded and unloaded her sled, and hundreds of times refused the proffered assistance of men to boost her sled up steep places, or turn back her sleigh when it would upset. A packer tells of offering to help her on one occasion, but seeing him with a pack on his back thanked him with the remark that he had enough of his own funeral. She was the only woman on either trail hauling by herself. Her dogs were the best trained of any; she never whipped, and fed them well and never over-worked them. She has them with her here

> and they may take her out over the ice. She stood the cold well, and the hard work—work that wore out many a man—did not seem to hurt her any. She deserves, and we know she will make the money she desires to relieve her home and make life comfortable in the future. Such a spirit as hers will always succeed.[4]

The Berry party, returning to their Eldorado claim in the spring of 1898, also did not wait for the ice to go out at Bennett. Tot Bush gives a good account of what dog sledding that spring was like.

> Clarence had spent much of his time scouting around the country looking for dogs to be used on the sleds. The main thing was to have them large and strong. He picked up some especially good ones. The front sled had a leader and eight dogs in double harness. Clarence manned this sled, which carried the provisions, dried fish for the dogs, and the chuck box, and so forth. On the second sled were the tent, the bedding, and bags and coats. If there was a stiff breeze, we used the sail and could ride on the sled ourselves; but we could not add our weight if the dogs were pulling.
>
> A great many lakes were grouped together, and we wove about, taking the one that looked safest for travel. . . . We were up at dawn, cooked our breakfast, and were on our way, trudging mile after mile and resting only when it became too dark to travel. It was bitterly cold. The lakes were frozen to a depth of six or eight feet, and snow was everywhere. Even with our running and our heavy stockings, our feet would get so cold at times that we could go no farther. All the dogs wore shoes made of several thicknesses of burlap tied around the ankle, but it would be only a short time until the sharp ice cut through the shoe to the foot. Then the dog would whine, and C. J. would say: "All right, old boy; I'll have you fixed up in no time." Then we would stop and put new shoes on any of them that needed it.[5]

At the end of each day, each person had a job to do in pitching camp—gathering firewood, cutting boughs for beds if there were some nearby, setting up the tents, cooking meals. As before, Ethel and Tot were in charge of the latter. At the end of one especially long and tiring day, when everyone was completely exhausted, the stove with supper on it suddenly melted through the ice and dumped everything. So Ethel and Tot had to put together yet another meal. But Tot was especially interested in the reactions of one of the men to the incident.

> I thought Charlie had a grand disposition, but that night he just went to pieces; he swore and refused to do his part. He said he

> wouldn't cook another meal if he never ate again, then left, and went to bed. His action was such a complete surprise to us that we looked at each other and couldn't speak. Then it struck us as funny, and we all began to laugh, telling each other that we didn't know he was like that. But Pa Berry made excuses for him, saying that the day had been long, and that he would be all right in the morning, and asking us not to give his bad temper any more thought. Poie (as we called Pa Berry) then offered to do Charlie's share. Poie was a dear, with the sweetest nature of anyone I ever knew. If ever you are in doubt about a friend, take him on a camping trip and you will never again have to wonder what your friend is like, for in three weeks you will know more about him than you learn in years.[6]

By the time Tot's party got to Lake LaBerge, the weather warmed and the ice thinned.

> We hugged the shore more closely each day. As there was constant danger of the ice breaking, we carried long poles; in case we should find thin ice the poles would prevent us from going through.
>
> The boys were mushing the dogs; as they had been going fast, Ethel and I were quite a distance back. When C. J. halted the sled to wait for us, he noticed that the runner was in water. Oh, boy! They used the whip on the dogs and worked like mad to keep the sleds moving. When Ethel felt the ice waving, she stopped, which was the worst thing she could do. Clarence looked back to see if we had noticed what was wrong with the sleds, and then Ethel called out: "Oh, Patty!" He couldn't leave the dogs and the sleds to come to her aid, and so he yelled: "For God's sake, Ethel, RUN!" When I heard that, I grabbed Ethel by the hand, and we ran! It was like being on the ocean, but the ice did not open. We thanked our stars when we got through with our outfits, but C. J. had been frightened for all of us. Had he left the sleds for a minute, they would all have been lost as, when the ice is in that condition and a dead weight rests on it, the ice gently opens, everything goes through, and the gap closes. That very afternoon a man with his dogs and his loaded sled went through in another place, but we did not hear of his misfortune until the next day.[7]

Most stampeders who arrived at Bennett in the winter and spring of 1897–98 did not go further on the ice. They chose instead to build a boat or raft and wait until they could float their outfits to Dawson. So a temporary community of several thousand bloomed at the lake's edge in the spring of 1898. Flora Shaw (Lugard), the English correspondent, described the scene at Bennett in her May 14 and 27 dispatches to the London *Times*. Her first view was from the Skagway/White Pass trail, approaching from the south.

> . . . Stretching before us for a distance of several miles was a glorious panorama of mountain and lake scenery. Lake Linderman was on our immediate left and a short distance beyond extended the long, narrow Lake Bennett, still covered with ice, closely encompassed by steep, rocky mountains. Between the two lakes and near the head of Bennett in particular was a considerable area of comparatively level land, and here, whichever way one looked, one saw a dense mass of white tents. The canvas city of Bennett contains a population of fully 5,000, and there are other large camps at the head of Linderman, in one direction, and as many points down the lakes, towards Dawson, in the other. . . .
>
> . . . The conditions of life in this temporary city of Bennett are, of course, novel and in many respects such as have never occurred before in any place in the world. To begin with, the whole population is under canvas. The majority of tents are formed into boat-building camps. But there is at least one business "street" known as Main-street. This is situated along the shore of the lake and extends for a distance of about half a mile. All along it, on either side, are canvas stores, hotels, saloons, doctors' and lawyers' offices, and other places of business. The hotels and saloons, of course, predominate, and occasionally at night these are pretty lively places. Nevertheless, there are surprisingly few drunken brawls, and for the general good order which prevails great credit is certainly due to the North-West Mounted Police and the British Columbia officials.[8]

Inga Sjolseth (Kolloen), the Norwegian immigrant from Seattle, visited Bennett several times while her party's boat was being built on Lake Lindeman. She was not so impressed as Flora Shaw was with Bennett's uniqueness, but she was responsive to the countryside and to the variety of people working at the lake's shore.

> *Sat., May 21* Sandvig and I went over to Bennett City. There was not much to see there except boats and scows. There are many different ways of building boats here. Some of them are more like boxes than boats. I saw a steamboat, quite large, which they had built in Bennett, and which shall go to Dawson City.
>
> . . . *Thurs., May 26* The weather is beautiful. They tell me it has been warmer today than at any time this spring. I was outside at

7–2. Tent city and boat building on the inlet between Lakes Lindeman and Bennett while ice is still thick on Lake Bennett in the background. (U. W., Hegg photog., 65)

7–3. Bennett from the mountainside looking southwest toward Lynn Canal on May 30, 1898. In the foreground are boats leaving on the newly thawed lake. In middle ground are tents at Bennett, with the inlet from Lake Lindeman to the right of center. The passes leading to Dyea are in the background at left of center. (U. W., Hegg photog., 549)

four o'clock this morning and it was so beautiful. The water was so still and clear that the high snow covered mountains with their green forests mirrored themselves in the water.

Fri., May 27 The weather is very changeable here. Today the wind has blown cold. This evening Sandvig and I were at Bennett City to buy a bread pan, but there was none to be found in the whole town. There are many people who are camping here now, and wherever one goes there are crowds. They are from all countries and wear many different kinds of clothes. I saw an old woman dressed in mens' clothing. She wore a yellow coat. Another woman who was dressed in black, walked arm in arm with her. It was a comical pair. Many turned and stared at them.[9]

Many stampeders' paths crossed at Bennett in the late spring of 1898 as the crowd waited for the thaw. Like Inga, Martha Munger Purdy (Black), the Chicagoan who was advised to "buck up" on Chilkoot Pass, remembered the steamboat at Bennett. She also told of meeting correspondent Flora Shaw.

One afternoon I was standing on the shore of Lake Bennett, watching the little group around the *Flora*, the first of the river steamboats to attempt the passage of Miles Canyon and the White Horse Rapids, when I saw a new woman, different from the usual type coming into the country. She was wearing a smart-looking raincape and a tweed hat, and, as she turned, I noted that she was not so pretty as clever-looking. I walked over and introduced myself. She told me her name, Flora Shaw, and that she had been sent to the Klondyke by the London *Times*, of which she was the colonial editor. She had allowed a month's time to make the trip from London, England, to Dawson, but it was taking her much longer than that. She asked me where I lived, and I replied, "Chicago—but I have lately come from Kansas, where my father owns a ranch."

"Do you know my brother? He lives in Eureka, Kansas." I did, as this was the nearest post office to Catalpa Knob, which proves, as has been done so often, that the world is really small.[10]

After careful observation of the activity at Bennett, Flora Shaw was skeptical that the boats or their navigators would prove adequate for the hard journey ahead. She nevertheless marveled at the optimism each stampeder had as to his or her own prospects for success.

. . . It is a moderate estimate that fully 3,000 boats will start for the Yukon River within another week or so.

How many will get there and how many will be lost on the way is another matter. On this point gloomy forebodings are freely indulged in by the "old timers" who have done the journey before, and it is only necessary to consider the circumstances for a moment to realize that these sinister predictions are not without foundation. To begin with, the great majority of the boats are being built by men who have never built boats in their lives before. They are to be manned and navigated, in the majority of cases, by men who have never been in a boat in their lives before. Add to these conditions the fact the waters to be navigated are exceedingly treacherous, and in some cases positively dangerous, on account of rapids and rocks, and it will readily be understood that when a mob of 2,000 or more boats, all struggling to get ahead of each other, start in a body for the same destination, there is likely to be trouble. This is admitted by almost everybody in the camp, though no one seems to think that he individually is taking any

7–4. Lake Bennett became navigable on May 30 in 1898. Here, those who have finished their boats head north, while those still building pause for a moment to watch before returning to their hurried construction. (U. W., Hegg photog., 228)

> risk. The whole situation strongly exemplifies the strange disposition of the mining enthusiast. Without any apparent object to be gained, men take the most desperate chances, and it is only the sacrifice of several lives through drowning that has convinced people that it is no longer safe to walk upon the ice.[11]

When the ice did go out on May 30, 1898, the thousands of people at Bennett melted away with it, following the flow to Dawson.

Frances Dorley (Gillis), a milliner and dressmaker from Seattle, was among those who left soon after the breakup. Frances had set out for the Klondike on her own, full of adventure, but against her parents' wishes. She teamed up with three men at Skagway, agreeing to do the cooking for the group using her own supplies, which they would replace when they arrived in Dawson. Here she describes her departure from Bennett and a near-tragic event which led to her meeting a future business partner.

> We bought a thirty-foot boat from a big Norwegian who was building himself a tidy fortune by constructing boats and selling them to those more impatient than he.
>
> On the morning we planned to leave Bennett, I was at the shore helping to load our outfit into the new boat. Suddenly the air was rent with frantic screams from a little woman who had been working a few feet from me. She was pointing excitedly out into the lake, and tears were streaming down her middle-aged cheeks. Her old flat-bottomed scow, on which was tied her skinny roan horse, was swirling out in the water. It had broken loose from its moorings and drifted away. The old horse, as distraught as his mistress, had set the scow to frenzied whirling by stamping his feet.
>
> Several men jumped quickly into their boats and rowed toward the scow. After many futile attempts they snubbed the unwieldy craft with hooks and ropes, and finally they succeeded in reuniting the nervous horse with the little woman. She thanked us all in heavy southern accents, and introduced herself as Mrs. Moore.
>
> We finished our preparations that day and launched our new boat. It carried us quickly and safely out of Lake Bennett, through the two miles of Caribou Crossing, and well on toward Lake Tagish before we met trouble. Our neat, expensive boat, made of soft cedar, was slowly coming apart at the seams. Quickly we put ashore, and while the men cut alder strips with which to reinforce our leaky boat, I baked a batch of bread over a campfire and caught some fish.[12]

Later, in Dawson, Frances Dorley and Mrs. Moore became partners in running a boarding house.

7–5. An early portrait of Frances Dorley (Gillis). (Y. A., Gillis Coll'n, 4480)

7–6. A group of men and women work on boats at what looks like the southwestern shore of Lake Bennett. (U. W. Child photog., 5)

7–7. Armada heading for the Klondike on a lake of the Yukon River. (U. W., Hegg photog., 562, Lake LaBerge)

When enterprising Belinda Mulrooney (Carbonneau), who packed furnishings for her hotel over White Pass with Soapy Smith's help, first entered the Yukon area in 1897, she too teamed up with some men at Bennett. And like Frances Dorley (Gillis), she bargained for transportation on their scow. A Dawson newspaper article about her explained the terms: "She kept the party supplied with fish and meat by aid of gun and rod, in the handling of which she is an expert."[13] Belinda, however, probably did no cooking, as that was a job which she hated.

It took the Norwegian immigrant Inga Sjolseth (Kolloen) and her friends, who built their own boat, a bit longer than Frances Dorley to get under way from Bennett. But they were ready to sail on June 14, 1898, over three months after leaving Seattle. Inga's diary entry for that day described their departure from their boat-building site on Lake Lindeman, just above King River Falls. Their goods were transported by wagon to Bennett, the boat having been taken down through the rapids, apparently without incident, a few days before.

> We got up this morning at 4 o'clock, had breakfast, and packed all our stuff together. We hired a team and everything was taken to Bennett. The boat took everything aboard, and we were ready to sail by noon, but the wind was so strong that we were unable to leave until late in the afternoon. The wind blew [only] a little after we left Bennett, and we could not go very far. So we landed at 9:30, made coffee and drank it. Then we straightened things up and slept under God's heaven. In the morning, when we had had breakfast, we sailed from there with all speed. The wind continued.[14]

The uncertain winds Inga mentions are typical of the Yukon lakes. Currents are seldom strong enough to carry fully loaded boats, so for the gold rush they had to be propelled either by oar or sail. The winds on these northern lakes can be strong, but they are so variable that one must be alert to avoid disaster. One moment you are rowing across a glass-smooth surface from which the surrounding mountains are doubled in clear reflections. Suddenly the winds freshen, and if caught by sails, your rate of progress increases tenfold. And just as quickly dark clouds can boil into the sky, bringing with them tumultuous winds that whip the lake into huge waves, threatening to swamp your boat or drive you off course and onto rocky banks or sandbars. The Yukon lakes, even when seemingly placid, command respect.

As the lakes' broad, deep waters flow northward into the confines of river channels, the current quickens, but it too can be quite variable. When banks widen to several hundred feet, the river is slow. When they narrow between cliffs, the water moves swiftly, and rapids and whirlpools form.

7–8. One boat has just shot through the hazardous Miles Canyon Rapids as another one enters the narrow channel. On the cliff at right are some observers. (U. W., Hegg photog., 711a)

The most hazardous rapids were about fifty miles from the north shore of Lake Bennett on a section of river between Marsh Lake and Lake LaBerge. Here the water dropped quickly while surging through unyielding canyon walls one hundred feet wide. The two sets of rip-roaring rapids which resulted—Miles Canyon and White Horse—were wild enough to make any traveler reconsider his or her direction.[15]

Many chose at this point to portage their goods laboriously around the rapids through mosquito-infested, heavily overgrown side trails and to let their crafts down by rope through the churning waters. (Later, wagons with wheels riding on log rails, somewhat like a train, would carry goods around Miles Canyon and White Horse Rapids.)

Others decided to take the quicker path right through the rapids, and many lost all they had. Travelers often remarked that goods could be seen strewn along the river banks for several miles below Miles Canyon and White Horse Rapids.

Among the first boats through the rapids in June of 1898 was the heavily loaded scow of Salome and Thomas Lippy, another of the successful mining couples on Eldorado Creek. The Lippys had been on the *Excelsior* when she arrived in San Francisco the previous summer and launched the stampede. They were among the wealthiest of those who had struck it

7–9. The tramway with log rails built to skirt Miles Canyon and White Horse Rapids. Mosquito netting on hats was a necessity in the warm months. (U. W., Hegg photog., 2160)

7–10. A scow takes on water in the White Horse Rapids, 1898. (U. W., A. Curtis photog., 46062)

rich on Eldorado Creek, but unlike many Klondike celebrities, when interviewed soon after her arrival Salome Lippy gave a very matter-of-fact account of her life on the claim.[16] Now the Lippys were returning to the Klondike with a new baby and provisions for another season of mining. Mont Hawthorne's account of their journey through the rapids is accurate in gist and excitement if not in detail.

> There was three men and a woman holding her baby on [Lippy's] scow. They hit a rock going into the canyon but they pulled it off with a rope and she didn't sink. It scared them bad. They got into the whirlpool and spun round and round. After they got out of there they hit a rock in Squaw Rapids. It wasn't a sharp one, so all it done was to throw them over and they bounced agin another one on the other side. That got them back on balance again and they made it just fine down around the point. But then they hit that big rock, "the boat buster." It stove in the bow of the scow but she was still afloat. Blamed if one of them fellows didn't let out a yell and jump right over into Whitehorse Rapids and drown his fool self. The scow come through the white water just fine. Lippy and the other man and the woman with the young one rode her down until she run aground on a sand bar down below and they come ashore without even getting their feet wet.[17]

Shortly after this incident, Sam Steele of the Northwest Mounted Police took charge and ordered that no women or children could be taken on boats through the rapids. Steele's order was imposed despite the fact that a number of women besides Salome Lippy and her infant had gone through safely. Some had even relished it.

One of these was Emma Kelly, Kansas's independent stampeder and correspondent who had challenged her packers on the Dyea/Chilkoot trail to keep up with her. She approached Whitehorse in October 1897 and said, "I wanted to see and experience this so-called danger, which men freely court, but which women may only read or hear of."[18] She had a thrilling ride through Miles Canyon, as her later description conveys.

> If you would know the sensation of a ride through the canon go you to the "chutes," shoot them, but imagine you are going twice as fast. Imagine, also, that on each side of you a perpendicular wall of rock, with sharp, angular projections, rises to a height of three hundred feet, and that under and around you are about forty sawmills running at full speed. Then, you may have a faint conception of a boat ride through Miles Canon.[19]

Then it was on to White Horse Rapids, after which Emma felt "a mad desire for its repetition," so she hiked to the top of the rapids and rode the

7–11. A fully loaded scow with rescue lines to shore, sinking below White Horse Rapids. (U. W., Hegg photog., 574)

7–12. Drying soaked goods after wrecking at White Horse Rapids, 1898. (U. W., Hegg photog., 719a)

wild waves again. The second time through she took an oar. About the experience she wrote, "I do not know when I ever enjoyed anything so much in my life."[20] And she loved the praise and recognition that she got for doing it. "I did not feel that I did a very dangerous thing, and at no time felt the least fear. I must confess, though, that the hearty congratulations and cheers of so many big, strong men, for 'the first woman who ever took an oar in a boat going through White Horse,' gave me a momentary feeling of elation that I had really accomplished a most dangerous feat."[21]

A number of women who arrived after June 1898, undoubtedly resented Steele's conventional sexism. Kate Rockwell, a free-spirited entertainer going into the Klondike in 1899 is an example. Rather than comply with the order, as most women apparently did, she set to work to disguise herself as a boy. She went through the rapids wearing overalls and a cap, despite the $100 potential fine if she were caught and found to be female.[22]

Martha Purdy (Black), the Chicagoan who had struggled over Chilkoot Pass, also defied Steele's order rather than walk the mosquito-infested five miles around the rapids. Her party hired a pilot to take them through the rapids, a job that several stampeders who were experienced boatmen took on for a short time to improve their finances before proceeding to Dawson.

> We sped through the canyon. There was a breath-taking interval before we were swept into the seething cauldron of the White Horse Rapids, where so many venturesome souls had lost their lives and outfits. Half-way through our steering oar broke with a crack like that of a pistol shot, above the roaring waters. For a tense moment the boat whirled half her length about in the current. Captain Spencer quickly seized another oar, calling coolly, "Never mind, boys! Let her go stern to." A second's hesitation and our lives would have paid the penalty. It took the boat only 26 minutes to get through.[23]

Yet another perspective on White Horse Rapids comes from Dr. Luella Day (McConnell), a physician from Chicago who also went in during the summer of 1898. Luella's privately published book, *The Tragedy of the Klondike,* is an unusually pessimistic account of the stampede and events at Dawson. Luella tells many stories of dishonesty and official graft, and White Horse rapids had its share.

When she got to the rapids, Luella walked around them. With her were her St. Bernard, Prince Napoleon, and a member of her party, Mrs. Wilson. As they walked along the shore by White Horse rapids, a boat carrying some of her former traveling companions was swamped by huge waves. Napoleon jumped in to save them, as he had done higher upstream to save Luella when her boat capsized. This time, however, to Luella's distress, Napoleon was caught in a whirlpool. After he was rescued, Luella analyzed the treacherous river. She found that there was a huge, submerged rock midstream which made two channels. The one to the left was not safe; the one to the right was better. She verified her theory by watching several boats go safely through the right channel, while those taking the left were nearly all wrecked. So she and Mrs. Wilson made a flag to direct boats into the right channel. But according to her, the men who were making money by piloting boats were not happy with her efforts.

> While we were [hanging the flag] the pilots were in consultation, evidently displeased at our action. Shortly thereafter I saw one of the pilots sneakingly pull the flag in so close to the bank that approaching boats could not see it till they were directly under it. Then again the Indian appeared in his canoe in the left channel. I was then satisfied in my own mind that the Indian was being used as a decoy to lure men to death who would not or could not pay the required fee to the pilots.[24]

So Luella hired a young man to guard the signal flag until the majority of stampeders who were coming in at the time were through the rapids.

White Horse to Hootalinqua

Beyond White Horse Rapids, the Yukon is continuously navigable by riverboat to its mouth, so commercial boats could be engaged for the remaining 460 miles to Dawson. But most rushers continued on their now well-tested crafts. Twenty-eight miles beyond Whitehorse is the thirty-mile long Lake LaBerge, made famous by Robert Service's poetic tale of the Yukon, "The Cremation of Sam McGee." The lake is also notorious to travelers for its capricious winds and weather. Inga Sjolseth (Kolloen), the Norwegian-born diarist from Seattle, described her first encounter with Native Americans on Lake LaBerge. They were apparently Southern Tutchone. She also tells of the perils of mosquitoes and storms on her three-day crossing of the lake, accomplished at the summer solstice when the Yukon is lit for almost twenty-four hours a day.

> *Sun., June 19, (1898)* . . . When we got to Lake LaBerge we caught a contrary wind and could go no farther. We are stopped now near an Indian village. These are strange people. Some of them have silver rings in their noses and feathers in their chin. They are so dirty and ragged that they are disgusting to look at. Sandvig and I went quite far up the mountain side and it was

> beautiful up there. S had his gun with him and intended to shoot wild game, but didn't see any.
>
> *Mon., June 20* Early today, in fact at 4 o'clock, we left our camp. There was a good wind for a little while, but it died rather soon, and we have not sailed very far today.
>
> *Tues., June 21* The mosquitoes were so bad that I could not sleep last night. We didn't set up our tent, but slept on the open ground. I got up and cooked rice at one o'clock and at 3:30 everyone got up and we sailed away. During this afternoon it has thundered and lightning has flashed. It has also rained terribly hard. At 5 o'clock we went ashore, set up our tent. We shall camp here until tomorrow.[25]

Francis Dorley (Gillis), the milliner and dressmaker from Seattle, remembered her somewhat more humorous encounter with a group of Native Americans at Lake LaBerge that summer. These too were probably from an Athapascan tribe.

> While we were still on Lebarge, a fierce storm blew down on us and we and other members of the desperate, crazy looking fleet were whipped ashore. As we secured our boat and clambered out, a group of Indians who were selling fresh trout to the hungry gold seekers came up to us. They clustered around me, examined me silently and thoroughly, reached out dirty hands occasionally to touch my clothing. Then they held a brief consultation. Finally the chief spokesman of the group stepped forward and addressed my companions: "Nice squaw. We like her. Which one of you does she belong to? We give you many fish, and even much money, if you leave her here with us."
>
> The men, all dumbfounded by this strange offer, were completely tongue-tied in their confusion and embarrassment. Finally Mr. Britton rose galantly to the occasion. Stepping between the Indians and me he murmured nervously that I was his squaw and not for sale. I, the "fine squaw," stood rooted to the spot, feeling more afraid than at any other time since leaving Seattle.[26]

Just after Lake LaBerge, the waters gather into a swift-flowing, curving river of crystal-clear water called Thirty-Mile. Here hazards came in the form of large rocks in midstream. One accident on Thirty-Mile proved disastrous, as this story from the *Klondike Nugget* recounts.

> Milley Lane started from Seattle last spring—we will call her Milley Lane because . . . we cannot advertise these people. She is a pretty faced girl of German antecedents and of good reputation. The party she came in with was well fixed and had several ladies

among their number. Milley was quite popular and proved herself adaptable and industrious. All went well as a marriage bell until Thirty-Mile river was reached. A rock—a wreck—outfits all lost—a wet shivering crowd on the bank with no provisions and hardly enough clothes on their backs to protect them from mosquitoes. Pitying passers by bring this girl of 18 summers to Dawson. With clothes all draggled and shabby and without a change of raiment she sought work for three long days. Pocket book and stomach empty, and employment refused, on the evening of the third day Milley found herself on the bank of the river with two courses open to her. She could either jump into the river or go to board with one of the madams in Dawson's Whitechapel. Long was the matter debated in her mind, but at last a youthful love of life triumphed. Within an hour the girl was seen bathed and dressed in satins and laces, her beauty enhanced by handsome apparel and the hair-dresser's art. Trail acquaintances were shocked, and when spoken to, the girl broke completely down and dissolved in tears. This is all true, happened last week and hardly forms an incident of one chapter of Dawson's history.[27]

Georgia White, the desperate mother from San Francisco whose trail experiences had already been depressing enough to challenge her sanity, also met with disaster at Thirty-Mile. Her diary entry for June 29 starts out optimistically while describing Lake LaBerge. But one can hear the break where the mornings' writing left off and that of the evening took up again. (Fort Selkirk, which she mentions as their destination, is where the Pelly River joins the Yukon. It is 224 miles beyond the lake, an impossible distance for one day. She may have meant Hootalinqua, which is about 30 miles beyond the lake.)

A beautiful morning after a windy night. Got up at half past six, had breakfast and are sailing down Lake LaBarge which is 31 miles long and very wide. We expect to reach Fort Selkirk today. A police station was not far from our camp and a policeman was on patrol. The mountains around here are not so high as on the first part of our trip. Well, I hardly know how to begin for on this bright day so much happened. We got out of Lake LaBarge all right and when about five miles down the river we ran on a rock and knocked the bottom out of our boat. Mr Karn, Voorhees and I were thrown in the river and current was very swift so we drifted for about ¾ of a mile. I was under for quite a while before Bert spied me, then when I came up amongst boxes, bags and trunks, he caught me by the back and helped me to catch hold of a box of coffee and we both hung to that. Mr. [Karn?] was drifting ahead of

7–13. Unloading goods from swamped boat on Thirty-Mile River, just downstream from Lake LaBerge. These rapids were eliminated when their dangerous rocks were blasted out in the spring of 1899. (U. W., A. Curtis photog., 46065)

7–14. Hootalinqua Post at the confluence of the Teslin and Yukon rivers, looking northeast. Most of the trees around such riverboat stations were cut to fuel the boats' steam boilers. (U. W., Hegg photog., 460)

us on another box until a small boat overtook us then I hung to that until I put ashore. Then I had to run back on the side of the river for about half a mile where Mr. Craig's camp was. They had a fire and made tea for us. In the meantime, Minna was on the boat on the rock and the boys threw a rope to her which she caught and was pulled in the boat and brought to camp where we were both undressed and wrapped in blankets and they dried our clothes and my dress fell in the bonfire and was burnt. We felt terrible. They said when I was in the water all I called was "Oh! my children!" The boys lost everything but a few sacks of provisions and we lost lots of our clothes and the boat is worthless. (There were three wrecks on that rock today.) There we met Mr. Sadlemire who sent his card up to me and wanted us to come down there. He lost everything on the same rock and the Stewart Company lost about $6,000 worth of goods and the Craigs lost $2,000. We ate a little supper and stayed with Mr. Craig. About 2 p.m. the boat got off the rocks and floated away. When we found it, it was miles down the river and one axe and a sack of bacon was in it.[28]

Even in a steamboat Thirty-Mile could be a challenge, as Susie Stewart recounts. Arora Susie Stewart had been born in Norway and apparently arrived in Skagway, March 17, 1898, as a single woman wondering "why I had come to such a Lord's Forgotten Country."[29] She married C. J. Stewart in Dyea, October 10, 1898, but it was nearly two years later before she joined him in Dawson. Susie did not have much good to say about her stay in Skagway as she waited for what she considered a safe way to get to Dawson. Her account illustrates not only how much things had changed in just two short years, but also the ironies the unpredictable Yukon can deliver.

The only good thing that happened during my stay in Skagway was the building of the White Pass Railroad. That road saved a lot of hardship and many lives. I staid in Skagway until the road was finished to Whitehorse. I waited two weeks [years?] for the first passenger train to go thru. The first passenger train left Skagway on the ninth of August, 1900, nine o'clock A.M. I stopped two days in Whitehorse. While there I went to see the canyon and the Whitehorse Rapids. It was a great sight. . . . Aug. 11th at 10 P.M. we started down the river on the old "Bonanza King," which looked more like a woodshed than a steamboat. We were sailing that night, the next day and the next night.

On the morning of the 13th I was up early and on the deck in the beautiful sunshine. We were on the "Thirty-Mile" River, there was not a breath of wind and the great sheet of water was just like glass. At seven-thirty I went in the dining hall, there were only three of us in at that time. Our breakfast had just been served. Suddenly the boat received a great jolt which threw me out of seat. The others rushed out but I sat on the floor and waited to see what else was going to happen. In a few minutes they returned together with the captain who told me that the boat had struck a submerged rock and was taking water badly. He said not to be alarmed, that there was a good landing not far distant, and he hoped that she would hold out until we got there. At nine o'clock A.M. we made the landing. She was so close to the bank that a gang plank was run ashore. The ship settled to the bottom a little on one side. There were five horses aboard and they were taken off first and they were willing and glad to get out. Next they all rushed out some freight. They hung hams and bacon on the trees and everything else that they could. An hour later the steamer "Baily" came along and took us all aboard and we started on the river again. We left behind the decorations and that was the last I ever saw of the old "Bonanza King."[30]

Stikine River Route

After the hazardous rocks of Thirty-Mile, the Yukon widens at Hootalinqua with the influx of the Teslin River from the southeast. The fine, gray glacial silt carried by the Teslin clouds the once crystal-clear water and causes a continuous hissing sound as it hits the sides and bottoms of boats—a sound which is ever-present from here downstream to Dawson. And with the addition of each new glacial stream, the hissing is amplified.

The flow of stampeders also widened at Hootalinqua. Those who had come in over the Stikine trail floated down the Teslin River to the Yukon. The Stikine trail was named after the Stikine River, which flows into the Inside Passage at Fort Wrangell, about two days' steamer travel north of Seattle. At Fort Wrangell stampeders could engage a river steamer for a 150-mile upstream journey, sometimes against heavy current, to Telegraph Creek. From there they packed another 150 miles overland to the head of Teslin Lake. There, as on the Edmonton, Chilkoot, and White Pass trails, they tediously constructed boats by hand to begin the 600 mile downstream journey to Dawson.

The Stikine River had led a previous generation of stampeders to its headwaters in the Rocky Mountains of British Columbia when gold was discovered in the Cassiar District in the 1870s. It was then that Nellie Cashman, a petite but daring Irish immigrant and professional miner,

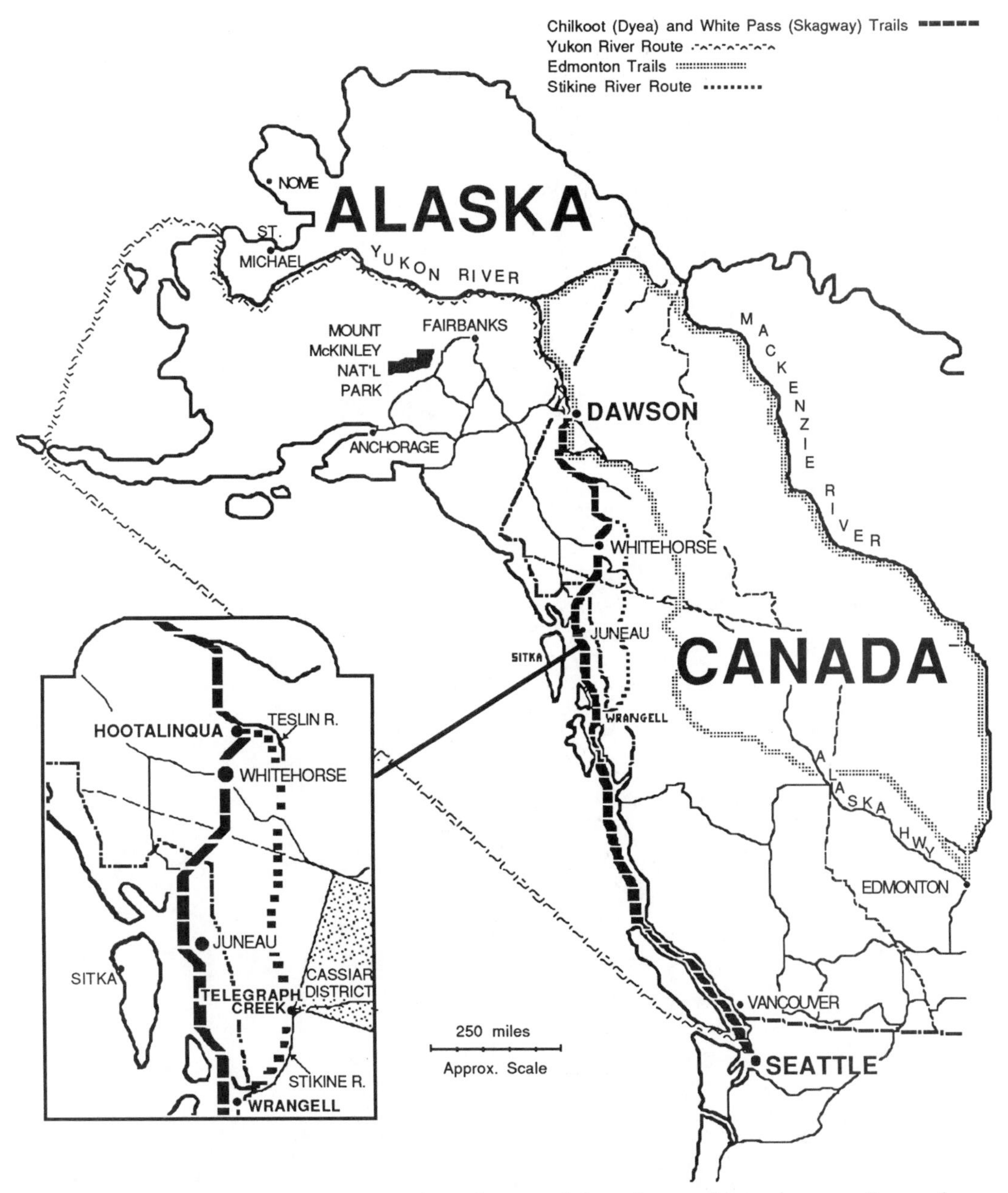

7–15. The Stikine River route to the Klondike is shown with large dots on this modern map. It passed close to the Cassiar Mining District of British Columbia. Since the headwaters of the Peace and Liard rivers are just across the Rocky Mountains to the east of the Cassiar, Edmonton trail stampeders sometimes converged with those on the Stikine route. (Modified from U.S. National Park Service map)

7–16. A painting of Nellie Cashman from the early 1880s. (A. H. S., 1847)

among other occupations, first rushed to the North Country to seek her fortune. Her selfless efforts to help cure fellow miners of scurvy in the winter of 1874–75[31] later earned her the epithet, "Angel of the Cassiar."

Twenty-some years later, Nellie was in Arizona when news of the Klondike strike reached her. By this time she was widely known in the West for her independence, adventuresome spirit, and compassion. Though later stories of Nellie gave her the aura of an angel, in reality she did not fit the conventional image of demure, sweet angeldom. One man describing her tempestuous spirit, called her "the wildest young lady I ever met."[32] As she got older, her admiring contemporaries referred to her as "Old Sour Dough" or "The Old Time Prospector."[33]

Though physically small, slender, and feminine, Nellie Cashman was also very sturdy and mentally tough. She had to be. Her restless spirit and impulsiveness took her throughout the western territories from the 1870s through the 1890s in search of gold and silver—Virginia City and Pioche, Nevada; Bisbee, Harshaw, Total Wreck, Tucson, Tombstone, Jerome, Prescott, Harqua Hala, Castle Dome, Yuma, and Globe, Arizona; Bodie, California; Kingston, New Mexico; Coeur d'Alene, Idaho; Cripple Creek, Colorado; the mountains of Montana and Wyoming; Baja California.[34] She even went to South Africa in search of diamonds.[35] One week she would be off to open a boarding house in a gold camp in Mexico. The next she was back in Arizona, her home base, thoroughly disenchanted with Mexico.[36]

Nellie was among the first settlers in Tombstone, Arizona. The early 1880s were Tombstone's heyday, when it had the reputation of being one of the toughest mining camps in the Southwest. These were the times of silver-lined prosperity, of the Earp brothers, Doc Halliday, the Clantons, and the shoot-out at the OK Corral. And in the midst of it all Nellie ran the Nevada Cash Store, selling groceries, shoes, and boots; the Arcade Restaurant; the Russ House; and the American Hotel and Restaurant. She raised money for numerous charities and civic projects, helping to build the Catholic Church and a hospital. And she reared her sister's five orphaned children.[37]

Though Nellie worked hard and made money in her various businesses and mining enterprises, she was so generous that she gave away whatever she or her nieces and nephews didn't really need. She was straight in her dealings with people, would put up with no nonsense, and had the courage of her convictions. She would do what she knew was fair and right, even if it meant going against the crowd. Nellie Cashman was of the stuff that legends are made.

By 1897, when the Klondike stampede started, Nellie's nieces and nephews were on their own. Nellie was nearly fifty years old,[38] but she was free and ready for adventure. A notice appeared in an Arizona newspaper on November 20, 1897: "Miss Nellie Cashman, one of the most favorably known women in Arizona, arrived from Yuma yesterday. Miss Nellie is preparing to organize a company for gold mining in Alaska. Her many friends in Arizona will wish her success, for during her twenty years residence in the territory, she has made several fortunes, all of which have gone to charity."[39]

Nellie set out for Dawson in the spring of 1898. This time she went in by way of the Chilkoot trail.[40]

The Stikine route was favored by Canadians early in the gold rush, because they thought they would have to pay duty if they were to land at the "American" ports of Dyea and Skagway. In 1897 the boundary between Canada and Alaska along the coast was disputed. Supposedly, it had been defined in 1827 by a treaty between Great Britain and Russia, the terms of which were transferred to the United States with the purchase of Alaska from Russia in 1867. However, interpretation of terms of the treaty differed between the countries. This meant that at various times in 1897, both Canada and the United States claimed Skagway and Dyea, as well as Bennett on the other side of the coastal mountains. The dispute was formally settled in 1903, but it had in fact been worked out by the events of the gold rush. United States citizens by far outnumbered Canadians in Dyea and Skagway. The Canadian mounted police set up and maintained entry points into territory claimed by "Canada" at the summits of Chilkoot and White Pass. The official border was finally drawn through the passes.[41]

Canadians did have the right of free passage on the Stikine River, so they would not have to pay duty on goods taken in by this route.[42] Another tangible advantage of the Stikine route was that the Teslin River entered the Yukon below the more formidable rapids encountered on the Chilkoot and White Pass branches of the Overland route.

Many parties entered the Klondike via the Stikine route, but it was not generally popular. The long overland section of the trail was more formidable than the thirty- to forty-mile treks of Chilkoot or White Pass trails, and it was not in good repair. One Skagway newspaper man, upon hearing a description of the Stikine route, wrote, "His account of the Stikeen trail would bring tears to the eyes of potatoes."[43]

By early 1898 the Canadian government signed a contract to put a railroad across the overland part of the Stikine trail. Information being distributed at that time confidently advised that it would be done by September 1898. What the reports did not anticipate was that financing the railroad would become a political issue in Canada. In fact, it was never built.

Among the stampeders on the Stikine route in the fall of 1897 was a Black couple, Charles and Lucile Hunter. Mrs. Hunter, still a teenager, was pregnant, and at Teslin she bore a daughter, whom she named Teslin. The Hunters stayed at nearby Atlin for Christmas. Then, rather than wait for the ice to break up, they pushed on to Dawson across the snow.[44]

If the race to the Klondike gold had been fierce before, at Hootalinqua Post competition was raised to a fever pitch as the Chilkoot and White Pass stampeders merged with those from the Stikine route. They all hurried on, down the last leg of the mad dash to Dawson.

DAWSON, AND THEN . . .

"What I wanted was liberty and opportunity"

FROM HOOTALINQUA to Dawson the Yukon River is comparatively safe. It winds calmly through thick, wild forest—forest that receded farther and farther from the water as it was consumed by the indifferent, hurrying crowds. They used the wood not so much for campfires as for the boilers of riverboats. With the stampede and subsequent development of the Yukon Territory, the number of paddle-wheel steamers increased dramatically. And each steamer devoured wood at the rate of two cords an hour when going upstream.

The general calm of this stretch of the Yukon River is interrupted only twice—by two rapids in close succession about 134 miles beyond Hootalinqua. Though formidable if the wrong channel is taken, Five Finger and Rink rapids are not nearly so dangerous as those farther upstream. Georgia White, the desperate mother from San Francisco who had barely survived the wrecking of her boat on Thirty-Mile River, was no doubt on edge as she approached Five Finger Rapids. The rapids get their name from the fact that four rock-faced islands split the Yukon into five channels.

> . . . went along until we reached Five Finger Rapids. They look dreadful. The rocks or fingers, are very large and sea gulls are flying around there waiting for victims. We unloaded a few hundred pounds and then Mr. Karns and Vorhees and the captain took the boat through O.K. We walked around which is about one mile through thick trees. We dislike very much to see Mr. Karn and [Bert?] go through but there was no help for it. We were very thankful when we met them on the trail. Then we reloaded and rowed down aways more. We camped and made supper, it being nearly 1 a.m. When we retired it was raining.[1]

8–1. Passengers and crew of the steamer *Eldorado* gather wood about 100 miles above Dawson. It was not unusual for passengers to help with this work since doing so meant getting under way again sooner. (U. W., A. Curtis photog., 27x)

8–2. Roping a steamer upstream through Five Finger Rapids as two stampeders on a scow approach the eastern channel at the same time. Observers on cliffs to right watch the action. (U. W., Hegg photog., 501)

Frances Dorley (Gillis), the milliner and dressmaker from Seattle who had teamed up as cook with three men in exchange for her passage, went through Five Finger Rapids in early summer of 1898, at about the same time as Georgia White. Her account not only captures the potential danger of the rapids, but also illustrates her low status in the party with regard to navigation decisions. Nevertheless, it is also apparent that Frances was far from being a passive passenger, and she rowed along with the rest when needed.

> I had bought a book in Skagway which told of the great dangers hidden in the foaming water that churned violently between the menacing rock spires from which the rapids derived their name. I suggested we follow the advice given in my book, and take the channel farthest to the west—supposedly the only safe course. The little Scot, McChord, hooted at my "book-learned knowledge," and shouted that I, being a woman, could know nothing about navigation. With true masculine loyalty, the other men sided with him.
>
> The argument continued until we were almost upon the rapids. As our shallow boat floundered helplessly in the churning water, the three men grudging conceded that I was probably right and that we'd better try to row across to the safer channel.
>
> We all grabbed oars and began rowing frantically with all our strength. We managed to force our way through the turbulent white foam until we were in midstream before we realized that we couldn't possibly reach the far side of the river. We were being drawn into the thundering rapids. We were all thoroughly frightened by this time, and it was a relief when Britton took command and told us to try to shoot through the center.
>
> I closed my eyes, said a little prayer, and tried to hold fast to my ebbing courage. Then we were shooting with brutal speed through the giant finger. Miraculously we passed through with our boat and our lives intact, albeit the former was cruelly battered and the latter sorely threatened.[2]

Modern river guidebooks advise downstream boaters to keep to the eastern or rightmost channel, the "giant finger" as Frances calls it, through Five Finger Rapids.

Between these last rapids and Dawson are 230 more picturesque miles. Except for sandbars, mosquitoes, and inclement weather, there was little for gold seekers to worry about but getting to Dawson as quickly as possible. The Yukon's volume of water increases with the addition, in turn, of the Pelly, White, and Stewart Rivers. Down the Pelly and Stewart

8–3. Camping at the mouth of the Stewart River on one of the placid stretches of the Yukon River, about 70 miles above Dawson. (U. W., Hegg photog., 2188a)

rivers came Edmonton trail stampeders who had gone overland or worked their way up rivers on the eastern slopes of the continental divide.

Occasionally, word would come of "colors" that had been discovered in creeks along the way, and stampeders would be diverted, at least temporarily, to stake a claim. Then usually they would continue to the more promising prospects of the Klondike. Finally, after weeks on the river, pulses would quicken at the words "There's Moosehide!" The big white scar left by a prehistoric landslide on the mountain behind Dawson can be seen for several miles upriver. It signaled that the stampeders had at last arrived at the fabulously rich Klondike Mining District.

At Dawson

What stampeders found when they reached Dawson depended on when they got there, what they were looking for, and how ambitious they were. For the earliest arrivals in the spring of 1897, there were plenty of opportunities. Dawson town lots were affordable, and there were still claims on the creeks to be located. Even if already staked, the full worth of a claim

8–4. Dawson waterfront in summer of 1897 or 1898. Wood and supplies for the expanding town lie scattered about. People may be waiting for a steamer. "Moosehide" shows on the mountainside at center. (A. H. L., Metcalf photog., PCA 34-80)

was not always realized. So many original locators sold out to newcomers for much less than the claim would eventually produce. In fact, possibilities in the spring of 1897 were wide-ranging for anyone with energy and/or imagination. One could buy a lot in Dawson or find a claim on a creek, build a cabin, fit it with furnishings made from the lumber of a disassembled boat, and get to work. There were even jobs available working for others in mining and in business.

By midsummer of 1897, however, spots in Dawson were considerably more dear. In only a few months, lot prices had soared to as high as $12,000.[3] Still, there were plenty of good mining and business opportunities.

By early 1898, the situation seemed bleaker. Cheechakos (as newcomers were called) found not only high prices in Dawson but very few promising creek claims that were unstaked. So, to the amusement of sourdoughs, the newcomers filed hillside or bench claims. Old timers' laughter turned to amazement when even the hills proved to be rich. They later learned the reason. The gold was in a riverbed all right but this river was so ancient that its bed had been raised by geological movements of the earth to levels above the modern creeks. When present-day creeks cut through these ancient riverbeds, they carried some of the valuable ore down into the modern beds. The older source of gold, however, lay yet in the uneroded hills. The cheechakos had found these ancient riverbeds. So there were still fortunes to be made for those who arrived before the main flow of the summer of 1898 stampede.

For those who got to Dawson after May of 1898, the situation had changed dramatically. Choice town lots then went for $40,000,[4] and with 30,000 people in town, there was practically no land available on which to pitch a tent or build a cabin. Many of the later arrivals simply stayed on the boats that had been their homes for several weeks and camped on the Dawson waterfront. Jobs were very hard to find. And men and women who came to the Klondike with the hope of living off the wealthy found instead thousands who were struggling just to keep themselves fed and warm. All the known gold-bearing creeks and hillsides had been staked, and claims were available for sale only at high prices.

Adding to the dreary scene were outbreaks of contagious diseases. Ironically, after weeks of hard but healthy life on the trails, crowds at Dawson exchanged colds, influenza, measles, pneumonia, smallpox, malaria, typhoid, and spinal meningitis. For those without money in the summer of 1898, Dawson could be a very discouraging place.

Early Arrivals

The women stampeders were no different from the rest. In general, great wealth was gained mainly by those arriving before 1898. Among them were Kate Mason Carmack, Salome Lippy, Ethel Bush Berry, Belinda Mulrooney (Carbonneau), Mae McKamish Meadows, Mary Ellen Galvin, and Ella Card.

KATE MASON CARMACK was, of course, really the first of the Klondike women. She, her husband George, and her brother Skookum Jim Mason had discovered the gold on Bonanza Creek. Like most other early Klondike women, she did very well, at least for a while.

After mining their rich Bonanza claims for two seasons, George, Kate, their daughter Graphie Gracie, and Skookum Jim all went outside in the fall of 1898 to celebrate and to visit George's family and friends.

While they were in Seattle, they managed to create a sensation. From the roof of their Seattle hotel, they amused themselves by throwing coins

8–5. George, Kate, and their daughter, Graphie Gracie, Carmack in about 1898. (*Post-Intelligencer,* Seattle)

to the crowds below, creating a near riot.[5] Kate, to keep from getting lost in the huge hotel, blazed a trail on molding and banisters with her hatchet.[6] And George rode around town in a carriage immodestly labeled, "George Carmack, Discoverer of the Gold in the Klondike."[7] The Carmacks bought property and talked of plans to build a yacht to sail to the world exposition at Paris.[8] In short, they made a splash.

But within a year, George and Kate's marriage was falling apart. In July of 1899 Kate was close to murdering George when authorities, roused by the commotion, broke into their hotel room. They found Kate on top of George on the floor, her hands around his neck. Kate was subdued only when she realized that she was in danger of being arrested.[9]

The following year, Kate sued George for divorce, charging desertion and infidelity and demanding her half of the $1.5 million estate. By that time George had returned to Dawson, leaving Kate and Graphie Gracie with his sister in California. In Dawson George met Marguerite Laimee, proprietress of a cigar store. They married October 30, 1900, in Olympia,

8–6. Kate Mason Carmack in Carcross, Yukon, shortly before she died. (U. W., Larson coll'n)

8–7. Upper Eldorado from the Lippy's claim at Number 16 in 1898. Thirty-two feet of gravel lay over the gold, all of it frozen solid. Wood fires were used to thaw the soil until it could be picked out and hoisted to the surface. There it was stored in great piles until warmer weather brought enough water to sluice the diggings. Wooden flumes crisscrossed the valley to bring the water to sluice boxes. (U. W., A. Curtis, 46168)

Washington, with George claiming that he and Kate were never really married.[10]

Kate never did get her $0.75 million. She returned to her home at Caribou Crossing (Carcross, Yukon), where she lived on a government pension. She was about 50 years old when she died March 29, 1920, of pneumonia and possibly tuberculosis.[11]

SALOME LIPPY and her husband Thomas had been in the North Country a couple of months when the Klondike discovery was made. They rushed to the creeks immediately. Thomas originally filed a claim high up on Eldorado Creek. But according to Pierre Berton[12] in his book *Klondike Fever* Salome convinced Thomas to change their claim to 16 Eldorado when it became available in the fall of 1896 because she wanted to live in a cabin for the winter. There was better timber lower down near the creek's junction with Bonanza, making cabin building easier, so they switched to Number 16. The change in location proved critical. Number 16 turned out

8–8. Clarence Berry shovels Eldorado pay dirt from the sluice box into the pan of his wife, Ethel Bush Berry. Also working on a pan in the midst of the clutter of the claim is Edna "Tot" Bush, Ethel's sister, who later married Henry Berry, Clarence's brother. (C. J. Bennett coll'n)

to be one of the richest on Eldorado, eventually yielding over $1.5 million. Claims higher up the creek were not so valuable. It was Salome and Thomas Lippy who staggered off the *Excelsior* in San Francisco with two-hundred pounds of gold. And it was the Lippys, returning to their claim in the spring of 1898, whose scow had been so battered in White Horse rapids. On the scow was Salome, with babe in arms, returning to Eldorado, where in the winter of 1896–97, despite the cabin, her adopted son had died. The Lippy's held their claim until 1903, after which they returned to Seattle with their great fortune and built a fine home. Eventually, the Great Depression made their investments worthless, however. When the Lippys died in the 1930s, they were bankrupt and nearly penniless.[13]

ETHEL BUSH BERRY and her husband C. J. were also very successful from their Eldorado claims. They had prospected without much luck in the Forty Mile District during the summer of 1896. When they learned of the discovery on Bonanza Creek, they both rushed to stake a claim.[14] And

8–9. The wealthy Bushes and Berrys after their return to California. Left to right are C. J., Ethel, Henry, and Edna "Tot." (C. J. Bennett coll'n)

they too were among the successful miners who set off the stampede in July of 1897 when they returned with their washings to the outside.

Ethel and C. J. worked as a team—C. J. directed the hired men in the digging and sluicing operations, and Ethel kept everybody fed, and doctored when necessary. In the warmer months, when the days were long, these were both round-the-clock jobs. But the claims at Eldorado 5 and 6 proved to be worth all the work.

EDNA "TOT" BUSH, Ethel's sister, who returned with them to Eldorado in the spring of 1898, also shared in their success. She spent the summer with them in their cabin on the claim and ended up taking responsibility for the household because Ethel got sick that summer, apparently with appendicitis. Tot's delightful book, *The Bushes and the Berrys*, has many lively, interesting descriptions of domestic life on the creek, as well as a humorous account of how she came to marry one of C. J.'s brothers, Henry Berry, who was mining on nearby Bonanza Creek.

The Berry family is unusual in that it is one of the few which maintained its Klondike wealth. After their Klondike success, C. J. and Ethel struck it rich again in the Fairbanks area in the early 1900s. And then their gold fortunes were reinvested in the "black gold" of California, oil. The Berry Oil Company is still a going concern in California.

BELINDA MULROONEY, the enterprising single woman who came in early via Chilkoot Pass, was one of the most colorful as well as the most successful of the Klondike women stampeders. She arrived in Dawson in the early summer of 1897.[15] From her roadhouse, hotel, mining interests, and mine management, she became one of the richest women in the Klondike.

When she first came in, Belinda brought not beans, bacon, and flour but goods which she estimated the early prospectors and their families would appreciate after a long winter of digging gold without contact with the outside—silks, hot water bottles, and cotton goods. She figured correctly and sold her merchandise to eager buyers at 600 percent profit. This she reinvested in a restaurant. At the same time she started building houses, as she explained many years later.

> There was nowhere then in Dawson for the newcomers to live and lumber was scarce as hens' teeth. "How are we going to shelter those people when they come in?" I said. "There'll be 30,000 of 'em by fall!" I started buying up all the small boats and rafts that were arriving, hired a crew of young fellows who had nothing to do and had 'em build cabins. I wasn't thinking of the money I'd make. We just had to shelter those people.[16]

As her profits accumulated, Belinda began looking for other investments. Dawson was booming, and there were undoubtedly opportunities there. But Dawson was not where the mining was. Belinda's innovation was to go to the heart of the diggings. So she left her Dawson businesses in the hands of caretakers and went to the richest gold-bearing region of the Klondike. Twelve miles from the Klondike River and fifteen miles southeast of Dawson, Bonanza and Eldorado creeks join, a mile and one-half above the original discovery claim. Here Belinda built a two-story log hotel and restaurant and called it the Grand Forks Hotel. Inadvertently, she started a town by the same name.

Belinda's hotel was a hit, and at first it had no competitors. No longer did tired and lonely miners have to go all the way into Dawson for some company, good food, or relaxation. They could meet at Miss Mulrooney's place. Her hotel became the center of social activity and conversation. And while her customers talked about the latest strikes, Belinda took note and later filed on any unclaimed ground. When discouraged miners decided to sell out and return to the comforts of home, Belinda would buy out their claims. By the end of 1897, she owned or had interests in at least five claims.[17] The number increased to even more the next year when she became partner and one-sixth stock holder in one of the wealthiest min-

8–10. Belinda Mulrooney's Grand Forks Hotel at the junction of Eldorado and Bonanza creeks in about 1899. Bonanza Creek would be in the foreground flowing right to left, if it were running. Eldorado is out of view to the right of the picture. The hotel became the center of the town of Grand Forks. (U. W., Child photog., 3)

8–11. A view of the kitchen of the Fair View Hotel. (U. W., Larss and Duclos, photog., 22)

ing companies in the territory, the Eldorado-Bonanza Quartz and Placer Mining Company.[18]

If Belinda Mulrooney's story ended here, it would be an interesting one. By the age of twenty-six she had amassed a fortune, and she had done it on her own. She could have settled back and lived off of her Klondike wealth for the rest of her life. But Belinda was having too much fun.

The next project to catch her imagination was to build and run the finest hotel in Dawson. Dawson townsite was being laid out and Belinda decided that a couple of lots overlooking the Yukon River were suitable for her grand hotel. So she bought the southeast corner of Princess Street and First Avenue, which paralleled the river. Then she planned and built the soon-to-be-famous Fair View Hotel.

Belinda ordered the finest of furnishings—real linen tablecloths and napkins, china, and silverware for the dining room; glass panes for all the windows; chandeliers—all to be shipped in and ready for the hotel's opening when the tide of newcomers arrived with the first boats of the summer of 1898. It was these furnishings that Belinda had transported, with the aid of Soapy Smith, across White Pass trail.

The Fair View opened on July 27, 1898, to a glowing review from the *Klondike Nugget.*

> It is well-located on high ground on Front Street, is three stories high, and has 30 guest rooms, besides ladies' parlor, gentlemen's smoking room, and [Turkish] bath. Every room is elegantly furnished and the hotel is fitted up with telephone, hot air for heating and for electric lights. The hotel will be run on both the American and European plan by manager J. K. Leaming, a gentleman who has been an hotel man for the past 20 years. Many of our readers will remember his being similarly engaged for the past six years in Los Angeles. The completion of the Fairview fills a long-felt want in Dawson. Miss Mulrooney is to be commended for her enterprise, for the hotel is by far the most pretentious structure now in Dawson.[19]

Mary E. Hitchcock, the Yukon River tourist, was an early visitor to the Fair View. She was obviously impressed by some of the things she observed in the new hotel, and maybe even a bit confounded by the juxtaposition of civilized refinements with frontier practicality. Her description is missing that glow of wonder and enjoyment that was evident in the *Nugget* article, but then Mary Hitchcock hadn't lived in Dawson and seen and felt what frontier life was really like. Her reference points were probably the fine eastern hotels with which she was familiar. At the same time, her report is more detailed about some aspects of the hotel—the food, the cook stove and the table linen—than was the local newspaper article.

We went first along the main street to a new hotel which was to be opened in the evening with a big dinner, followed by a dance. The house, built of wood, and three stories high, quite towered above the tents and cabins of its neighbours. The only entrance that was finished was through the new and elaborately furnished barroom, within whose walls many a sad history will probably be recorded during the coming year, as we are told that "the liquor business here is bigger pay than the richest mine," and that "even the smallest barroom realises between five hundred and a thousand dollars a night." Separated by a hallway from this saloon is the dining-room, beautifully clean, table covered with damask, and even napkins (something unusual in this part of the world) at each place. The menu, beginning with "oyster cocktails," caused us to open our eyes wide with astonishment, after all that the papers have told us of the starvation about Dawson. We next visited the kitchen adjoining, where there was a stove that would have gladdened the heart of any cook at home. The chef was said to be from Marchand's, of San Francisco. The proprietress explained to us that she had sent for chairs, which had arrived without legs, they having been left on the dock at St. Michaels, one of the inconveniences that one had to bear through the negligence of transportation companies, so she had carpenters at fifteen dollars a day manufacturing new legs.

On the second floor, a long, narrow hall separated rooms that were about double the size of an ordinary cabin on shipboard. Each room contained a primitive wooden bedstead, but there was no space for wardrobe, closet, or dressing table. Evidently the pride of the hostess's heart was centered in Brussels carpets and lace curtains, to which she called our attention as having been introduced into Dawson for the first time. The price of one of these tiny rooms was six dollars and a half a day, food five dollars extra, or two dollars a meal. On the third floor the carpenters were busy preparing for the evening dance, after which the large hall was to be partitioned off in to small rooms, at five dollars a day each, providing that the sojourn of the guest should be at least of one month's duration, otherwise terms to be increased accordingly. We were cordially invited to return for the dinner at 10 p.m., and also for the dance. Noticing that there were no panes of glass in the windows, which were simply covered with cheesecloth, we asked what happened in case of rain, and were told that it very rarely rained, but that when it did there would probably not be sufficient to do any damage. Glass also had been ordered, but, as usual, it was impossible to tell when or by what steamer it would arrive.[20]

The "hostess," of course, was Belinda Mulrooney. To the old-timers in Dawson, the fact that the Fair View had central heating was a wonder. Belinda recalled in a 1908 newspaper interview how she had managed the problem of heating her hotel: "A furnace was not even thought of; but I had one built of a coal oil tank."[21] She ran the hotel for about one year, "until it justified my investment by more than paying for itself. I then leased it for $1,250 a month, having cleared in twelve months about $30,000."[22] She eventually sold the Fair View in 1904.

But Belinda didn't sit still while running her hotel. She was constantly working out partnerships and business deals, investing in mines, and lending money. She helped to start the telephone company for Dawson and Grand Forks, and the water company that at last brought safe drinking water to Dawson.

She also met and married Charles Eugene Carbonneau, a charming man and brilliant conversationalist,[23] who called himself a Count, some say fraudulently.[24] Although later stories would have them first meeting in the spring of 1900 when the dapper Carbonneau appeared at the Fair View to sell champagne,[25] in fact Charles had been in the Territory as early as August of 1898, and Belinda had business dealings with him by at least June of 1899. She apparently lent him money, for she was given a mortgage by the Anglo French Klondyke Syndicate, of which Charles was manager, on two of their claims on Eldorado and Bonanza Creeks.[26] They married October 1, 1900, in what was undoubtedly one of the most elaborate weddings in Dawson.

For a few years after their wedding, Charles and Belinda were in the Klondike during the summer, supervising work on the mining interests. During the winter they went to Paris where they lived in grand style along with Belinda's three younger sisters and a brother. In the summers of 1903 and 1904 Belinda returned to Dawson alone, while Charles stayed in France. She took over management of the Gold Run Mining Company, at that time probably the largest mining operation in the Klondike. She and Charles were partners in the company along with several others. The Gold Run Mining Company had lost money at first, but Belinda, with her fine managerial skills, was expected to turn things around. And apparently she did.

But then came setbacks. Charles was charged in Paris with embezzlement and fraud.[27] The Gold Run Mining Company sued Belinda for letting lays on the mining properties, saying she was not authorized to do so.[28] And then the bank which held the mortgage on the Company sued it for the debt.[29] When Belinda left Dawson for Fairbanks in the fall of 1904, she was probably just walking out on the whole mess. She never returned to Dawson.

According to one source,[30] Charles was responsible for losing all of Belinda's money. In any case, in 1906 Belinda divorced Charles,[31] a drastic

8–12. Charles Eugene and Belinda Mulrooney Carbonneau shortly after their wedding in 1900. (Seattle *Times* photog.)

8–13. Belinda Mulrooney Carbonneau in the garden of her castle in Yakima in the early 1920s. (Gue coll'n, Gladys Wilfong and Marjorie Lambert)

action for the times, and especially for a Catholic, as Belinda was. Thereafter, she claimed to be a widow.

Then Belinda and her sister Margaret applied the Mulrooney magic to making money in the Fairbanks gold rush. As she had done in Dawson, Belinda went directly to the producing creeks and set up a business—this time the Dome City Bank. And once again Belinda amassed a fortune. With this Fairbanks-area money she moved to Yakima, Washington. There she build a castle and surrounded herself with her family. Her youngest brother, sisters, and nephews lived with her at the castle, and she built a home nearby for her parents.

Belinda lived at her castle until the late 1920s, but for several of her later years there, her fortune was exhausted. So she rented the castle for income.[32] Eventually she moved to Seattle where she died in 1967 at the age of ninety-five.[33]

MAE MCKAMISH MEADOWS, the high-spirited Californian who survived the Sheep Camp flood, probably arrived in Dawson in late November 1897. According to one newspaper account,[34] she was worth $100,000 by mid-April the following spring, having invested in business, real estate, and mining. Referred to by her stage name, Mae Melbourne, the story says she also grubstaked some successful gold-seekers. In exchange for lending them money so that they could prospect, she received one-half of anything

8–14. Mae McKamish Meadows as she appeared in an 1898 newspaper story about her Klondike success. (*Klondike News,* April 1, 1898.)

they found. Grubstaking was common practice in the Klondike and was a way for both men and women to be "miners" without actually getting out with a pick and shovel themselves. The amount of Mae's worth mentioned in the article, however, might be exaggerated. The story was probably written by her flamboyant husband, Arizona Charley Meadows, who also served as General Manager for that special edition newspaper.

Mae and Charley stayed in Dawson for a number of years and continued to make headlines. They built the Palace Grand Theater in 1899, one of the few gold rush buildings still standing in Dawson today. It wouldn't be if one of Charley's ambitious schemes had materialized. In December 1899, only six months after opening the Palace Grand, Meadows awarded a contract to float the entire three-story theater to Nome in the spring when the Yukon River became free of ice. Apparently it was already clear to Mae and Charley that Dawson was on the decline and that the new gold discovery at Nome was where the action was. Elaborate details were published in the *Klondike Nugget* as to just how the removal of the theater was to be accomplished. The story included the note that passengers and gear riding

8–15. Mae McKamish Meadows in Yuma, Arizona, early 1900s. (Jean King coll'n)

8–16. Mae McKamish Meadows with Buffalo Bill Cody in Yuma, Arizona, about 1913. (Jean King coll'n)

on the first floor would serve as ballast to keep the theater from capsizing off of its rafts.[35] But the plan never was executed, and the Palace Grand stayed in Dawson, though it was later named the Savoy Theater.

The Palace Grand was the scene for plays, dances, and various social functions, as well as regularly scheduled stage shows. Charley and Mae were featured in some of these. In one melodrama Charley rode on stage on his horse. He also gave demonstrations of his pistol skills by shooting glass balls that Mae held in her hands or had resting in her hair. That act was dropped when Charley accidentally shot Mae in the thumb one night.[36]

By the fall of 1901 Mae and Charley sold the theater and left Dawson. Charley's idea at that time was to take a gang of cowboys to Tiburon Island in the Gulf of California to subdue the "cannibalistic" natives there and develop the island as a resort.[37] But they settled in Yuma, Arizona, and took up their position as one of the ten richest families in the area.[38]

Eventually, in 1926, Mae and Charley separated. Mae went to Santa Monica, California, where she lived until she was 88 years old. She died there December 21, 1960.[39]

8–17. Mary Ellen and Pat Galvin on their claim at 5 above discovery on Bonanza Creek. This early photo gives an idea of what the creeks looked like before mining operations had completely transformed the valleys. (*Klondike Nugget*, April 1, 1898)

MARY ELLEN GALVIN, who had likened the Yukon River to an ice house blown up with dynamite, like Salome Lippy and Ethel Berry, was an early arrival on the creeks. She had been with her husband, Pat, a hardware merchant and stove maker, at Fort Cudahy (Forty Mile River) when the Klondike strike was made.[40] They bought interests in several different claims on Bonanza and Eldorado creeks, which proved very valuable.

In addition, Pat had a knack for shrewd business investments. That together with boundless energy, a hearty constitution, and Mary Ellen's help had netted them $1.5 million by April of 1898.[41] But by 1899 their businesses had floundered. They were bankrupt when they left the Yukon.[42]

ELLA CARD, whose infant daughter died at Lake Lindeman, came on to Dawson with her husband Fred in the summer of 1897. In July, she had another baby, Freddy.[43] So Ella Card had been six months pregnant when she crossed Chilkoot Pass. She and Fred each operated a hotel in Dawson for several years and apparently did quite well. But at some point they

8–18. Ella Card (Clark) when she lived in Fairbanks. (A. H. L., Wickersham coll'n, PCA 277-11-39)

divorced. Then in September of 1904 disaster struck again. The hotel Cecil, which Ella had been managing and in which she had invested all her capital, burned to the ground. Ella stayed in the building until the last minute, heroically making sure that all of her guests were out. Then she found herself trapped on the second floor and was forced to leap from a window. Fortunately, she was not seriously injured by the fall, but she was financially wiped out. There was no insurance.[44]

After the fire, Ella went to newly-founded Fairbanks, Alaska, scene of the latest gold rush. There she opened a restaurant, the Cecil Cafe, and became a longtime resident.

Later Arrivals

Of the women who came into the Klondike in 1898, many who were determined but adaptable were able to make a good living, though none became fabulously rich.

LUCILLE HUNTER and her husband Charles, who came in by way of the Stikine route, located a bench claim on Bonanza Creek on February 26, 1898.[45] They spent the rest of their lives in the Yukon, running successful claims in both the Klondike and Mayo areas. Their daughter, Teslin, grew up on the creeks and undoubtedly received much attention from homesick

8–19. Inga Sjolseth Kolloen, at center, with husband Henry and an unidentified woman at the Jo Jo Hotel near Dawson, about 1898. (U. W., Wolfe photog., 1)

miners. When grown, Teslin married a fisherman in Seattle, but she died soon after. After Charles' death in the early 1930s, Lucille kept three claims in the Klondike and a silver claim in the Mayo district represented on her own. In order to keep a claim, a miner must work on it each year. Lucille did the required work, and since she had no car, she walked the 140 miles between her various mines. She died at the age of 94 in Whitehorse, June 10, 1972.[46]

Inga Sjolseth, the Norwegian immigrant from Seattle, supported herself for over a year doing cooking and possibly laundry in Dawson. On July 15, 1899, she married Henry Kolloen, whom she had known from Seattle and who had come in via Chilkoot trail at the same time as Inga. Together they ran a successful roadhouse, the Jo Jo Hotel at Gold Run Creek, about fifteen miles from Dawson.

According to their son Erling, when Inga and Henry left the Yukon in 1901, they "came Out with enough 'dust' to take a three-year vacation in their native Norway, to buy a home in Seattle, and to lose a few thousand in the Scandinavian-American Bank which had to close its doors."[47] Though they were never really wealthy, they did spend the rest of their lives comfortably in Seattle.

8–20. Frances Dorley Gillis on the porch of her home in Dawson, with Dr. A. J. Gillis and possibly Dr. A. Thompson. (Y. A., Gillis coll'n, 4461)

FRANCES DORLEY, the milliner and dressmaker from Seattle, arrived safely in Dawson July 5, 1898, after almost two months on the trail. Her first impression of Dawson was of its confusion and activity. And as she recalled many years later, she jumped right in. "Almost at once I liked the muddy, gold-crazy settlement of shacks and tents. I spent the next three months reassembling my stores of food and equipment and getting acquainted with my colorful surroundings and robust neighbors. By October I had secured a cabin at the junction of Eldorado and Bonanza Creeks, the two richest gold-bearing streams in the region."[48]

Frances's cabin was in Belinda Mulrooney's now prospering Grand Forks. She turned her cabin into a roadhouse where she offered homemade grub, her specialty being baked beans. "Butter was shipped up from the States in wooden buckets which contained five pounds each, at about five dollars a pound. I saved all the empty butter tubs I could find and filled them with baked beans. To the hungry men, starved for good, well cooked food, they were worth many a dig into their gold pokes."[49]

The following spring Frances moved back to Dawson, where she was reacquainted with Mrs. Moore, the Southern woman whose raft and horse had broken away at Bennett. They became good friends and together opened the very successful Professional Men's Boarding House.

Frances met her husband-to-be, Dr. A. J. Gillis, a dentist, at the end of 1899 when she consulted him about a tooth she had chipped on a Christmas hazelnut. They were married in 1902 and enjoyed together the outdoor life that the Yukon so abundantly offers.

In 1917 Frances bore a daughter, and shortly after, in 1919, the Gillis family moved from Dawson. Frances eventually returned to Seattle after her husband's death in 1929. But as a seventy-year-old in 1947, Frances Dorley Gillis remembered her days in the Yukon as the happiest of her life.[50]

NELLIE CASHMAN, the fifty-year-old Arizona miner who had been "Angel of the Cassiar," reached Dawson in the late spring or early summer of 1898, having come in over Chilkoot trail. As she had done in mining camps throughout the West, Nellie opened a restaurant, The Can-Can, in Dawson. She also ran a small store in the basement of the Hotel Donovan.[51] It was in front of this store that John P. Clum, an old friend from Tombstone days, photographed Nellie—somewhat stouter now than twenty years earlier.

At the same time, Nellie began prospecting and acquiring claims. But strangely, most of her Klondike mining interests were beset by trouble and litigation.

In July, for $1,500 cash, she bought a bench claim on Little Skookum, a tributary of the fabulously rich Bonanza Creek.[52] It had been located February 2, 1898, by Mrs. Frances Johndrew (Gendreau). Now, when people located claims on the Klondike creeks, they didn't have surveyors along. They just measured off one hundred feet square of ground and put up stakes. When surveyors came in later, they often found that errors had been made. Extra, unclaimed ground might turn up, in which case a fractional claim could then be made. Or, as happened on Little Skookum, there could be just too many claims for the available ground.[53]

The dispute came to a head when Nellie hired a man and sent him to start working her claim. The workman was stopped by the owner of the neighboring claim, Russel, who said it was his ground. When the argument was settled, Nellie found she had lost half of her claim, and it turned out to be the only half that had any gold. Russel took $20,000 out of that piece of ground, while Nellie got only the expense of working her own half.[54]

In October, Nellie fell back on another business in which she had experience and opened a small cafe in Dawson, the Cassiar Restaurant. Early the following year she staked a claim herself on Monte Cristo Hill above Bonanza Creek, but again the claim was disputed. This time Nellie declared that her stakes had been tampered with, leaving her with only

8–21. Nellie Cashman stands in front of her store in Dawson, June 23, 1898. (A. H. S., Clum photog., 1134)

a fractional claim, which she applied to have granted to her. Her fractional grant was approved, but then a suit brought by her neighbors resulted in a survey which showed that even the fraction was part of some other claims. Her Monte Cristo Hill claim was simply surveyed out of existence.[55]

Nellie had always felt that the dispute with Russel had been settled unfairly. After the Monte Cristo Hill fiasco, she petitioned Gold Commissioner Senkler for some compensation for her good-faith losses. Eventually, in 1901 and after much petitioning, the authorities agreed that errors had been made and granted Nellie ten claims on Soap Creek, a tributary of Gold Bottom. But these compensation claims never proved very valuable.[56]

Nellie's richest claim was one she bought in partnership with five others in 1901. It was 19 below discovery on Bonanza Creek. Reports were that she made as much as $100,000 from that claim, eventually buying out her partners. But she put all of it into charity or prospecting and ended up with very little.[57]

In addition to her mining interests, Nellie involved herself from the very start with the charitable works that had become her trademark. The June 30, 1898 Yukon *Midnight Sun* reported that she had just returned from a very successful trip to the creeks where she collected donations for Dawson's hospital. Built by Jesuit missionary Father William Judge, St. Mary's Hospital was supported by community donations. It was no sooner completed than it was filled to capacity. Nellie made sure that everybody had a chance to show their community spirit by donating.

Throughout her life Nellie obviously enjoyed and perhaps preferred associating with men. When she heard that many Klondike miners went to the saloons at night just for company, she transformed a part of her store under the Donovan into a kind of social hall. It became known as "The Prospectors' Haven of Retreat" where any miner, usually male, could stop in for a free cigar or a cup of coffee and some conversation and companionship.[58] John Clum recalled, "She was generous to a fault, always helping some worthy—but hard-up miners."[59] She herself was often broke, but Alex McDonald and Jim McNamee, who had each made great fortunes in the Klondike, would "supply her with sufficient gold to put her back on easy street, . . . [and they] did so without any thought of reimbursement. In fact, whatever they gave to Nellie was considered an indirect donation to charity, for they were quite sure that sooner or later, their gifts and her winnings would all be disbursed to the needy and afflicted, to churches and hospitals, and, therefore, it was only a matter of time until Nellie would be broke again, and it would be up to them to provide her with another 'stake.' "[60]

Nellie Cashman's later epithet, Angle of the Cassiar, captured the almost magical powers which grew to surround Nellie as the years passed. One old miner from the Cassiar, when on his death bed, expressed the firm belief, "If Nellie Cashman were only here, I'd get well."[61] Mae Field, who had come to Dawson as a newlywed on June 20, 1898, recalled another episode which seemed to confirm Nellie's almost miraculous powers.

> [Nellie Cashman] was an angel if there ever was one! We all loved her. One year there was a terrible fire in Dawson, and everyone thought the whole town was doomed to go up in flames. Then Nellie Cashman asked for a bowl of Holy Water, which she threw into the raging inferno. To everyone's utter amazement, the wind changed. Almost instantly the flames were swept in the opposite direction, and the greater part of the town was saved.[62]

In 1904 Nellie left Dawson, saying, "I've got to go down to the Tanana and feed the boys. Some of them will be hungry like in all new camps and as usual I will be there to feed them on coffee and pork and beans."[63] She headed for Fairbanks, the town that sprang up in the Tanana district, and continued to live the life that legends are made of. From Fairbanks she moved further north, above the Arctic Circle, to the remote Koyukuk District. There she remained until 1924, making occasional trips into Fairbanks for supplies and to the outside to raise money and visit her family.

It was on these later trips that the now-legendary Nellie was interviewed about the motivations and attitudes that had led to her success. When asked why she had never married, she first laughed, then said more seriously, "I have always been too busy to talk about such things."[64] In answer to another interviewer's similar question, she elaborated, with a twinkle in her eye. "Men are a nuisance anyhow, now aren't they? . . . Men—why, child, they're just boys grown up. I've nursed them, embalmed them, fed and scolded them, acted as mother confessor and fought my own with them and you have to treat them just like boys."[65]

Nellie went on to reveal the real affection and admiration that she had for her comrades.

> It takes real folks to live by themselves in the lands of the north. Of course, there are some rascals everywhere, but up north, there is a kindly feeling toward humans and a sense of fair play that one doesn't find here, where men cut each other's business to hack and call it "competition." It takes the solitude of frozen nights with the howl of dogs for company, the glistening fairness of days when nature reaches out and loves you, she's so beautiful, to bring out the soul of folks. Banging trolley cars, honking cars, clubs for catty women and false standards of living won't do it.[66]

Nellie's disdain for the "civilized" world came out again, along with some philosophy, in a newspaper report that announced her return to Fairbanks in 1919, on her way to Koyukuk.

> After giving the East country the "once over" and taking two or three looks at the short skirts of the girls who had no tundra to mush over and the high heels they affected and on which no woman could ever accomplish the trail and arrive at the messhouse in time for meals, she had a dentist fix her teeth and came right away from there, back to the everwas land, which is so far away from California that California will never amount to much.
>
> Nellie Cashman knows that there are no hardships or privations in the world—that they are all in one's mind. She knows that things are never so bad that they could not be far worse, and when she is thrown into association with people who have no pursuit but the avoidance or the treatment of influenza, and no pastime except the complaining or talking about the high prices

8–22. Nellie Cashman in about 1923 when she was in her seventies. (A. H. S., N. H. Rose photog., 2090)

> of the straightform corsets, it makes her peevish—she has seen and known life, and that baby stuff is not for her. She listened to their complaints of ill health, watched their eyes inflame and the tears run from them; saw them using innumerable pocket handkerchiefs to keep their noses clean, and then, disgustedly, advised them to get out of there and come to Alaska where life was life and not one nose wiping exhibition after another, and hit the trail for the North, leaving them to follow her advice and herself, or be miserable ever after.[67]

Nellie was asked about her recommended appointment as Deputy U.S. Marshal in an interview in 1921. She said she considered herself better fitted to enforce the law than anybody else in the district, then she good-humoredly explained her own notions of law enforcement.

> No, not the enforcement of the two-gun man. I wouldn't think of using force on anybody, particularly those boys up there. You see, they look on me as a sort of mother and they wouldn't think of doing anything wrong while I was around. I've been all through Alaska dozens of times but I've never been troubled by bad men. There isn't a man in Alaska who doesn't take off his hat whenever he meets me—and they always stop swearing when I come round, too. I wouldn't have any trouble in keeping order, because everybody's orderly when I'm round, anyway![68]

Nellie's trips to the outside also made headlines because even as a seventy-year-old, she often covered hundreds of miles across the snow by dog sled. When she came out in January of 1924 to raise capital for her latest venture, the Associated Press carried the following article.

> Miss Nellie Cashman, of slight figure and worn by years of prospecting and mining in the north, fully maintained her reputation of being the champion woman musher of the north in the opinion of pioneers here, when she came to Seward recently to take a steamship to the states. To reach Seward, Miss Cashman mushed, that is to say, part of the time she ran behind a dog sled and part of the time rode by standing on the runners, 750 miles in 17 days. In the 17 days, with a good dog team and the lightning fast ice that precedes the heavy snows, she traveled along the Yukon and Tanana Rivers from Koyukuk, Alaska to Nenana whence she rode the cushions of the government's Alaska railroad.[69]

Upon returning to Alaska from her trip outside, however, Nellie became ill with pneumonia, from which she never fully recovered. But

Nellie even approached death as an adventure. She said, "You never quite know what's going to happen next, or when your time will come to cash in your checks. It all adds interest and variety to life."[70] She died January 4, 1925, in Victoria as she was making her way to her family in Arizona to convalesce.[71]

MARTHA MUNGER PURDY (BLACK), the Chicagoan who had trekked the Chilkoot with her brother after her husband had decided to go to Hawaii, was somewhat unusual in her success. Although she arrived rather late (August 5, 1898), she ended up doing quite well. Perhaps part of the credit goes to the money she and her brother had brought with them, which helped them to get established during the first year. But a large part is no doubt due to her unusual determination as well.

Martha spent the final months of her pregnancy and her first winter in Dawson in a small cabin her party had built across the Klondike River from Dawson. When she bore her third son, Lyman, in January 1899, she was assisted by two old prospectors, one with a hook for a hand, who took the precaution to wrap it with cotton to make it soft for the baby.[72] When Martha's father arrived the following July to take her and the baby home, she had already decided she was not returning to her husband Will. (They divorced in about 1901.) Martha wanted to make a new life in the Yukon, working a claim she had staked at Excelsior Creek on her way into Dawson. But she was convinced to go to her parents' home, at least until the claim could be proven. Ironically, the confining comfort of Catalpa Knob, Kansas, proved one of the hardest environments to bear, and she fell into a depression.

> I was only 33 years old. So many years stretched ahead of me—interminable—uninteresting. Somehow the mainspring of my life had snapped, my zest for life, for adventure was no more. I lost interest in my clothes—which to a woman of my temperament who loves pretty things is one of the last props. . . . What I wanted was not shelter and safety, but liberty and opportunity.[73]

Fortunately, Martha's Excelsior claim proved rich. She returned to Dawson in 1900 and started and managed a very successful sawmill. In 1902 while working on some mill business, she consulted a young lawyer in Dawson and thus met George Black. Martha recalled one of their early conversations: "Mr. Black commented on conduct of mine to which he objected. I told him, 'I'm paying you for legal advice . . . if I need any other kind I will ask for it!' He told me later he could have spanked me!"[74]

But apparently George wasn't too put off by Martha's feistiness. He soon asked her to marry him. Two years later, on August 1, 1904, they were wed, beginning a long partnership that was fully intermingled with

8–23. Martha Munger Purdy Black and George Black on a Klondike River fishing trip in about 1915. (Y. A., M. L. Black coll'n, 3263)

8–24. Martha with husband George to her right and son Lyman at her feet on board the troopship *SS Canada* in 1916. (Y. A., National Archives of Canada coll'n, 381)

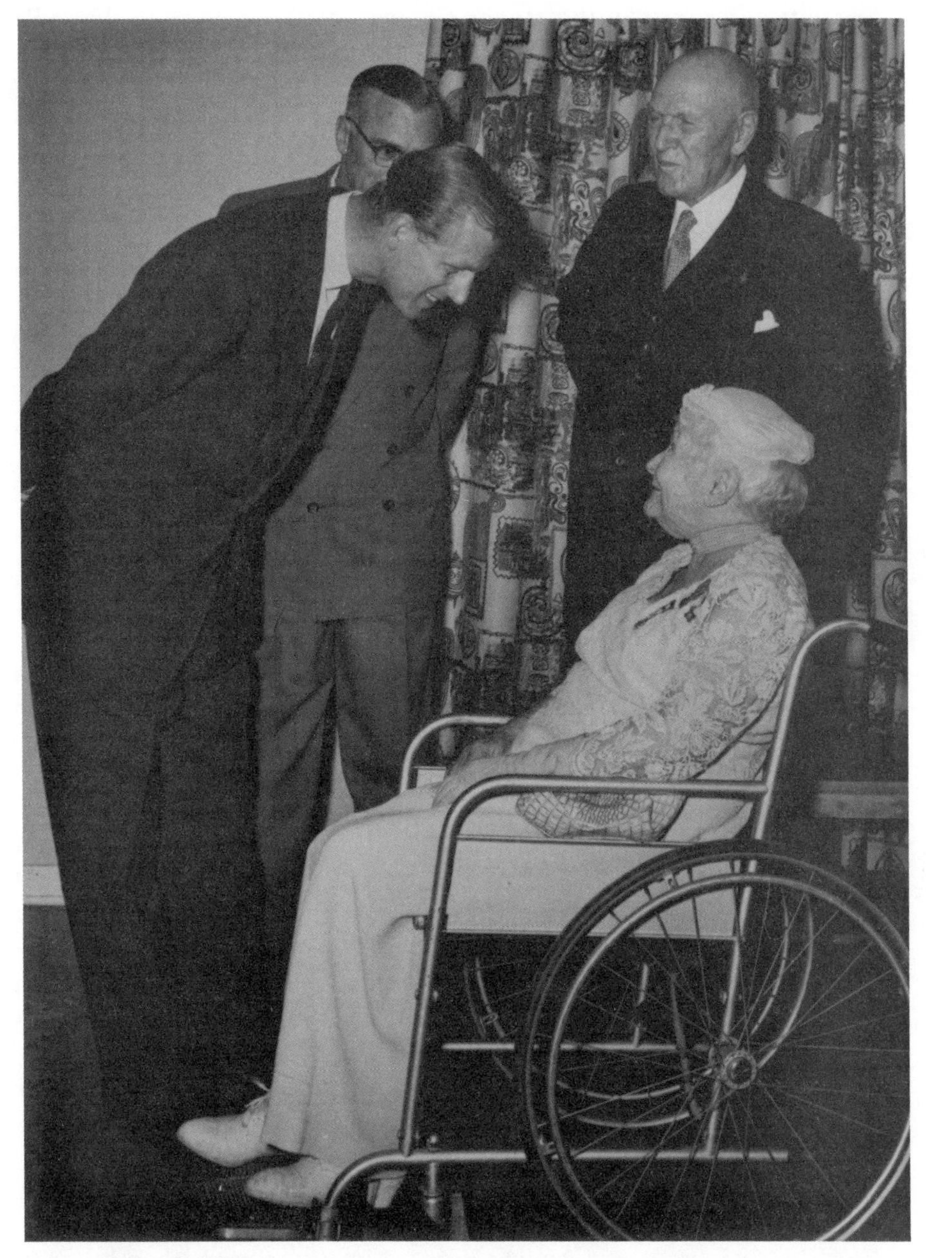

8–25. Martha and George Black meet with the then Duke of Edinburgh at Whitehorse in 1954. (Y. A., MacBride Museum coll'n, 3598)

the political fortunes of the Yukon. George was first elected to the Yukon Council and then in 1912 was appointed Commissioner of the Yukon Territory.

Martha not only supported George in both positions but became fully involved in the politics and interests of the territory. As chatelaine of Government House, she opened the residence to all the Yukon constituency and thoroughly indulged her love of flowers by developing the Government House gardens. It was also during these years that Martha was commissioned to prepare exhibits of wildflowers which she collected for the Canadian Pacific Railway.

When George resigned the Commissionership in 1916 to enlist for World War I and organize the Yukon Infantry, Martha wasn't going to be left behind. She sailed with him and the company to England. There she acted as war correspondent for the Dawson *News* and the Whitehorse *Star* and looked after the Yukon volunteers whenever they were in London.

Martha's account of her life, *My Ninety Years,* is exceptionally discreet, but her comments on some of her experiences in England, give a glimmer of her independence of spirit which fit so well with her life in the Yukon. She was certainly sophisticated and much taken with royalty and pomp and circumstance. And yet she could not accept the observance of form without feeling, and she chafed, though with good humor, at the artificial restrictions placed on women. Upon being led to the remote, grilled Ladies' Gallery at the House of Commons she comments, "I told some of the members later that this was my first visit to a country where men were so frightened of women that they had to keep them behind bars!"[75]

After the war, Martha and George went first to Vancouver, then to Ottawa when George was elected the Yukon representative to the Canadian House of Commons, of which he became Speaker in 1930. During these years Canada's relationship with Great Britain changed from that of a colony to a member of the Commonwealth. When George was forced to resign from the House of Commons due to illness in 1935, Martha was elected in his stead, a Conservative party victory in what was otherwise a Liberal landslide. At age sixty-nine, she was the second woman to serve in the House of Commons, but she considered her position only an interim one until George was well again. He did resume his seat in the early 1940s. In 1942 the Blacks bought a house in Whitehorse, then becoming the center of Yukon commerce because of its position on the newly constructed Alaskan Highway, which bypassed Dawson. When George left office in 1949, the Blacks retired to Whitehorse, where Martha died on October 31, 1957, at age 91.[76]

Another latecomer who did quite well was the extremely popular entertainer known as CAD WILSON. Cad was brought to Dawson in the fall of

8–26. Cad Wilson, third from left, entertains in her Dawson apartment with "Diamond Tooth" Gertie Lovejoy. (A. H. L., Wickersham coll'n, PCA 277-1-190)

1898 by the highest salary ever paid to an entertainer on the West coast.[77] But that was not her first trip to the Yukon. In 1894 she had hiked over Chilkoot trail as secretary for Veazie Wilson's photographic expedition down the Yukon River. When the gold rush began, she wrote a series of interesting adventure travelogs combining descriptions from 1894 with those of current conditions. Her articles are full of humor and appreciation for the North Country. And the photos accompanying them often show her in hamming poses.[78] In 1898 then, Cad Wilson was no stranger to the Yukon or to adventure. And she was an instant success in Dawson.

Though not particularly beautiful or even a very good singer, Cad was a hit because she was a professional who knew her audience. Her previous experience had been in classical theater in New York where she was known as Esther Lyons.[79] In Dawson, she combined good humor with what newspapers at the time referred to as a "ginger." Her performances were to overflow crowds, and her appreciative audience bruised her with tossed nuggets and coins at the end of the act.

However, she did miscalculate on one occasion. It was a benefit performance where "ladies" were present. (Women had, of course, always been at Cad's performances prior to this, but they were performers and women who earned their money by dancing with the men. By the standards of the day, they were of a different type.) What had been appreciated by the men of her usual audience was considered too audacious for the more delicate taste of ladies. Though there is no description of exactly what Cad did, other than sing some of her most favored songs, the *Klondike Nugget's* editorial censure, "Was It Clean," leaves no doubt as to her effect on at least some of the audience.

> [Cad's] act caused ladies to reach for their wraps and many and severe were the comments at the conclusion. They felt they had been inveigled to a charity benefit under false pretenses. . . . Cad Wilson made no friends by her acts that night. Her audacity called out applause in the rear of the hall, but the ladies in front hung their heads and their escorts wished they had never brought them. . . . Words are hardly strong enough to express our condemnation of anyone who deliberately and premeditatively insults the better part of an audience.[80]

Apparently Cad was no less popular for the indiscretion, however. She continued to perform to sellout crowds, and she joined in the laughter over parodies by some male performers of her hit song, "She's Such a Nice Girl, Too." When she left Dawson after less than a year, she had $26,000 in money and nuggets.[81] And she carried a token of at least the male population's esteem—a huge golden belt, which even the *Nugget* had to admit was of "almost barbaric magnificence."[82]

> The pin of the fastener is behind a gold pan the size of a dollar containing a perfect model of a windlass coiled with a golden rope and bearing a golden bucket, which has evidently just increased the size of a golden dump. On each side of the gold pan is a pick and shovel, binding with golden cord an occasional nugget. The brooch part is nearly six inches long and from this is suspended five nugget chains bearing those precious trinkets with which the ladies love to burden themselves. The articles carried are a ladies handsome gold watch, with the monogram "C. W." on the back and the front rim studded with golddust; memorandum tablets with back and front of heavy gold; bon-bon box of gold, the center of the lid carrying a handsome quartz nugget, with lettering around the circumference in colored gold, "Klondike, N. W. T."; an extension gold pencil in a gold-dust studded case; a gold-dust studded perfume bottle of pure gold. The gold pan above bears the inscription of "Dawson" also in raised colored letters.[83]

8–27. The Anderson children with their dog Bruce in the Yukon. Left to right are Dewy, Ethel, and Clay. (U. W., Hegg photog., 475)

Cad Wilson a.k.a. Esther Lyons had indeed done quite well for herself. But as with many other Klondike successes, she did not maintain her great wealth. Her fortune was spent long before she died on October 27, 1938, in Elizabeth, New Jersey, where she had worked occasionally as a practical nurse. She was 74 years old.[84]

EMMA (MRS. P. B.) ANDERSON, five-year-old ETHEL ANDERSON (BECKER)'s mother, who came in with the children in the fall of 1898, might also appear to be an exception. However, she does fit the pattern of success if husbands and wives are considered as a team. Her husband, Peter B. Anderson, had come into the Klondike in late 1897 or early 1898.[85] From Ethel's account, they lived and worked on 16 Eldorado, Lippy's claim, where her father was a carpenter, and when they left the Yukon in 1902, they "literally dripped nuggets."[86]

As an adult Ethel fondly remembered her early years in the Yukon and wrote articles and books about the gold rush and Klondike mining. She also wrote proudly of her mother's ingenuity in coping with rearing three young children on the creeks—keeping them healthy when diseases swept the country, finding and inventing amusements, avoiding the hazards of heavy machinery and open mining shafts, keeping everyone fed and clothed, and helping her children enjoy the natural wonders of the creeks.[87] Ethel Anderson Becker died at age 77 in Seattle in 1970.[88]

LUELLA DAY, the physician from Chicago, also did well for herself, though she did not end up being fabulously rich. She arrived in Dawson about mid-June of 1898. For a while she practiced medicine. But then the Canadian physicians organized and required enforcement of the law which said that all doctors must be registered. To register, one must pass a licensing exam, and to take the exam, a certificate for completion of a four-year medical course had to be presented. The problem was, there was no examination procedure yet set up in Dawson, and several recognized medical colleges in the United States had only three-year courses available. So Luella had to stop calling herself a doctor. Instead she hung out her sign as a nurse. In the meantime, people were dying in Dawson for lack of medical attention.[89]

On September 26, 1898, Luella married Edward McConnell. She had seen Ed soon after she arrived in Dawson and was immediately taken with his good looks, intelligent conversation, and civilized manner. He was a successful miner, and a steamboat owner, and he owned and operated the ferry across the Klondike River.[90] Together, Luella and Ed operated the Melbourne Hotel in Dawson. As she describes in her book, *The Tragedy of the Klondike,* however, Luella soon became embroiled in bitter disputes with officials of the territorial government, who she was convinced were crooks.

While Martha Munger Black used the height of discretion in writing her memoirs, *My Ninety Years,* Luella Day McConnell's book is a no-holds-barred indictment of the early government of the Yukon. In this respect, she presents an unusual perspective on the gold rush and the life in Dawson. While it is obvious that Luella thoroughly distrusted the government officials, it is difficult now to assess the validity of her charges, which include bribery, blackmail, harrassment, extortion, conspiracy, fraud, and murder. In fact, she describes how on two different occasions officials made attempts to murder her with arsenic.[91]

After the second of these attacks, Luella did manage to escape from Dawson with some of the money she had accumulated from the Melbourne. (In her book, Luella does not mention that she and Ed McConnell were married, nor does she explain what happened to him at the time of her escape.) But she was frozen in for the winter on her way down the Yukon. At that point she heard of the Fairbanks strike. So she went to Fairbanks in April of 1904, and stayed there for a couple of months before going, by way of Nome and San Francisco, to her home in St. Augustine, Florida.[92]

The other late arrivals of 1898 whose trail stories have been told here returned to the outside with little gold to show for their efforts.

GEORGIA WHITE, the desperate mother from San Francisco, arrived in Dawson on July 8, 1898, but found it very dull. It was August 4 before she found work in a laundry, where she earned $5 for a hard day's labor. She was ill and dispirited a good part of the time and eventually left Dawson by downriver steamer on September 6 to return to San Francisco, really no better off financially than when she had left.[93] And yet, Georgia was far from defeated or dissatisfied by the North Country. She divorced her husband and returned to Alaska in 1900. After another return to the lower states, she went to Alaska again in 1902 and married Frank Mills, who had been aboard the *Australia* with her when she first left San Francisco on the stampede. She died in 1904.[94]

NETTIE HOVEN, the plucky, beautiful young woman who had worked her way from New York in 1898 on Mrs. Gould's widow ship, ran a roadhouse on Hunker Creek for about a year. In August of 1899 she married an athletic young man from San Francisco named Fred Thoerner.[95] John F. Mellen, her lover from New York who had threatened to kill her, did follow her to Dawson. When Nettie appealed to the courts for protection, Mellen was warned to stay away. He agreed to do so.[96]

EMILY CRAIG, the stalwart Danish immigrant from Chicago who had been two years on the Edmonton trail with her husband, arrived in Dawson in August of 1899 to find the Klondike rush already over. There were no paying claims to be had. Mr. Craig left February 19, 1900, for Nome, where he was able to earn a living as a carpenter at the newly discovered gold fields there. Though Emily supported herself for a few months as a cook, she was broke when she left Dawson on May 28, 1900, to join her husband. "I left the Joyal [Joyelle?] Hotel, after having spent a very nice and pleasant winter, as far as circumstances would allow, that is, having to work for my living, and earning my passage money to go to Nome, and wishing I was back in Chicago, and wondering why I came. After all, it is long to be remembered."[97]

The Craigs stayed in Nome until 1909, where A. C. worked as a carpenter. In 1915 they moved to Anchorage when the Alaska Railroad between Seward and Fairbanks was under construction. When A. C. died in 1928, Emily became a nurse at the Railroad hospital, where she met Dr. J. H. Romig. They married in 1937 and eventually retired to Colorado Springs. When Dr. Romig died, Emily moved to Seattle, where her adopted daughter lived. She died there May 7, 1957, at the age of 86.[98]

Other Kinds of Gold

These stories of the Klondike gold rush of 1897–98 are not complete. Except for the interview material, they are based only on written records in public archives. And for the most part, they include only first-person accounts. Many women who participated in the stampede did not leave written accounts of their experiences. Or if they did, their reports are not public. Notable among the missing voices are those of prostitutes, for example.

Even the stories that were recorded have not all survived the one-hundred or so years since the rush. Some have been lost or destroyed. Some may simply be buried, waiting to be "mined" from letters in hometown newspapers of the 1890s or from family records.

The stories here are also incomplete in that they are often excerpts from longer narratives. The original sources listed in the Reference section can be consulted for additional valuable, interesting material.

What the selected written reports gathered here do tell us is of the variety of educated, mostly Caucasian women who joined the stampede and participated in the early mineral exploitation of northwest America. Their stories vary widely from tales of adventure to tales of woe. But generally they seem to have one common theme: Although not all of the Klondike women were financially successful, most seemed to feel richly rewarded by their gold rush experiences. And for those who longed for it, the North Country usually did provide the liberty to live their lives fully and the opportunity to make their own ways in the world.

NOTES

Introduction

1. San Francisco *Chronicle,* July 21, 1897, p. 1.
2. Ibid., p. 2.
3. Ibid., p. 1.
4. When referring to women at the time of the gold rush, I use their names at that time. Later married names are indicated in parentheses.
5. Seattle *Post-Intelligencer,* March 18, 1898, p. 10.
6. Ibid.
7. Selma [Calif.] *Irrigator,* reprinted from the Sutter Creek *Record,* July 24, 1897, p. 3.
8. Selma *Irrigator,* April 9, 1898, p. 6.
9. Spellings of proper and place names can vary widely. For example, "Stikine" was also "Stikeen," and "Skagway" could also be "Skaguay." I use the currently accepted versions in the text. Spellings in quoted items are those of the source.

Chapter 1

1. E. A. Becker, "Little Girl in the Klondike Gold Fields," *Alaska Sportsman,* November 1962, p. 22.
2. E. F. Larson, interview, September 24, 1979, p. 1; and J. E. Feero obituary, Skagway *Daily Alaskan,* December 8, 1898, p. 1.
3. Larson, interview, September 17, 1979, pp. 18, 33; and Larson obituary, Tacoma *News Tribune,* January 28, 1981, p. C–8.
4. Larson, interview, September 17, 1979, p. 3, and September 24, 1979, p. 4.
5. Larson, interview, September 24, 1979, p. 1.
6. Larson, interview, September 17, 1979, p. 1, 18.
7. See, for example, W. Ogilvie, *Early Days on the Yukon* (Ottawa: Thorburn Abbott, 1913), p. 129ff.
8. See, for example, J. R. Little, "Squaw Kate," *Alaska Life,* March 1943, pp. 17–18; J. D. Hartman, letter to E. A. Becker, September 29, 1942, University of Washington Alaska Biography, N979.819; "Kate Carmack," *The Pathfinder,* August 1920, p. 22; H. W. Jones, *Women Who Braved the Far North,* vol. 1., San Diego: Grossmont Press, 1976; and

E. A. Becker, "Discoverer of the Klondike," *Seattle Post-Intelligencer,* Pictorial Review, August 18, 1963, pp. 10–12. A similar account was given by Evelyn Day, in interview of July 13, 1982.

9. See George W. Carmack, *My Experiences in the Yukon,* privately published, 1933.
10. Larson, interview, September 24, 1979, p. 2.
11. Larson, interview, October 23, 1979, p. 8.
12. Larson, interview, September 17, 1979, p. 1.
13. Mae McKamish Meadows, letter, August 24, 1897, in Santa Cruz [Calif.] *Daily Sentinel,* September 8, 1897, p. 3.
14. Larson, interview, September 17, 1979, p. 1.
15. Ibid., p. 2.
16. Although Edith says in her interview of September 17, 1979 (p. 2), that she arrived in Skagway on the *Al-ki* on August 19, 1897, this is probably inaccurate. The *Al-ki* left Tacoma August 4 and arrived in Skagway on August 11 (see T. Adney, *The Klondike Stampede of 1897–1898* [New York: Harper, 1899], p. 44). By August 18 it had returned to Seattle and left Tacoma again August 19. So if Edith did go to Skagway in August, she probably landed about August 10 or August 23, 1897. Another source, H. Clifford, *The Skagway Story* (Anchorage: Alaska Northwest, 1975), reports that the Feero family arrived in October. I have not been able to verify either set of dates.
17. Larson, interview, September 24, 1979, p. 1.
18. Larson, interview, September 17, 1979, p. 2.
19. Ibid., p. 3.
20. E. A. Becker, "Little Girl in the Klondike Goldfields," *Alaska Sportsman,* November 1962, pp. 23–24.
21. For those living in the Northwest, "in" referred to the Yukon and Alaska, while all other parts of the world were "out."

Chapter 2

1. Stacia T. Barnes Rickert, Pioneer Women of Alaska Scrapbook, Clara Rust Collection, University of Alaska, Fairbanks.
2. Tappan Adney, *The Klondike Stampede of 1897–1898* (New York: Harper, 1899), pp. 14–15.
3. J. G. MacGregor, *The Klondike Rush through Edmonton, 1897–1898* (Toronto: McClelland and Stewart, 1970), pp. 245, 263. I have added one other woman, Isadore Fix, to MacGregor's tally. She is mentioned in the text (p. 166) but is omitted from the list on p. 263. Her marital status is not given. I am assuming that the Mrs. Sam Brown listed by MacGregor is the same as Mrs. Braund whom Emily Craig Romig reports meeting in *A Pioneer Woman in Alaska.* Although some details

about them vary, their stories are very similar. Mrs. Braund was traveling with her husband, and according to E. C. Romig, bore a baby at Fort McPherson on January 6, 1899. They got as far as Fort Yukon, then went to St. Michaels instead of Dawson.

4. Fresno *Daily Evening Expositor*, September 23, 1897, p. 1.
5. Ibid., August 4, 1897, p. 1; September 4, 1897, p. 8.
6. Ibid., September 23, 1897, p. 1.
7. Ibid.
8. J. Wilbur Cate, in Fresno *Daily Evening Expositor*, October 28, 1897, p. 1.
9. MacGregor, op. cit., pp. 201ff.
10. Ibid., p. 5.
11. Juneau *Alaska Searchlight*, March 27, 1897, p. 3.
12. MacGregor, op. cit., p. 236.
13. Edmonton *Bulletin*, September 5, 1898.
14. Elizabeth Page, *Wild Horses and Gold* (New York: Farrar and Rinehart, 1932), p. xi. This is an excellent, novelized account of one party's struggles to reach the Klondike by the Edmonton overland trail.
15. MacGregor, op. cit., pp. 233–36. For an account of one who died after arriving in Dawson, see W. Galpin, "Who Is to Blame?" in A. R. Crane, *Smiles and Tears from the Klondyke* (New York: Doxey's, 1901) pp. 119ff.
16. MacGregor, op. cit., p. 173.
17. E. Craig Romig, *A Pioneer Woman in Alaska* (Caldwell, Idaho: Caxton, 1948), p. 20.
18. Ibid., p. 23.
19. Ibid., p. 19.
20. Ibid., pp. 24–26.
21. Ibid., p. 28.
22. Ibid.
23. Ibid., p. 30.
24. Ibid., pp. 26–27.
25. Ibid., pp. 34–35.
26. Ibid., pp. 38–39.
27. Ibid., p. 42.
28. Ibid., pp. 50–51.
29. Ibid., p. 59.
30. Ibid., pp.59–60.
31. Ibid., p. 63.
32. Ibid., pp. 65–66.
33. Ibid., p. 65.
34. Ibid., pp. 72–73.
35. Joseph Ladue, *Klondyke Facts* (New York: American Technical Book Co., 1897), pp. 29–30.

36. Romig, op. cit., p.74.
37. Ibid., p. 75.
38. Ibid., pp. 81–82.
39. Ibid., pp. 88–89.
40. Ibid., pp. 91–92.
41. Ibid., p. 104.
42. MacGregor, op. cit., p. 95.
43. Romig, op. cit., p. 82.
44. Ibid.
45. MacGregor, op. cit., p. 135, quoting A. D. Stewart in Hamilton Ontario *Herald.*
46. Ibid., pp. 236ff. A. D. Stewart died March 13, 1899, on the Peel River of scurvy.
47. Ibid., pp. 168ff.

Chapter 3

1. M. E. Hitchcock, *Two Women in the Klondike* (New York: Putnam, 1899), p. 27.
2. Ibid., p. 30.
3. See, e.g., J. Lynch's critical comments in *Three Years in the Klondike* (Chicago: Lakeside, Donnelley, 1967) p. 18f, and those of P. Berton in *Klondike Fever* (New York: Knopf, 1958), p. 316ff.
4. Hitchcock, op. cit., pp. 22–23.
5. Tappan Adney, *The Klondike Stampede of 1897–1898* (New York: Harper, 1899), p. 192.
6. Hitchcock, op. cit., p. 76.
7. Ibid., pp. 456–57.
8. Hannah S. Gould organized the Women's Clondyke Expedition of five hundred women from New York, many of them widows. Several of them sailed on the steamship *City of Columbia* around the Horn to Seattle, while others came cross-country. At the bon voyage party for the *Columbia,* Mrs. Gould spoke of a number of enterprises the expedition would undertake upon reaching Dawson, including establishing a library, church, recreation hall, restaurant, and hospital. See, e.g., Seattle *Post-Intelligencer,* May 1, 1898; ibid., May 4, 1898; Berton, op. cit., p. 135.
9. Dawson *Klondike Nugget,* July 30, 1898, p. 1.
10. Ibid., August 23, 1899, p. 3. It could be that Nettie Hoven is an alias for Kate McRae, who was single, worked as a stewardess on the *Columbia,* and arrived with the Gould party in Seattle. Seattle *Post-Intelligencer,* May 1, 1898.

Chapter 4

1. Martha Louise Black, *My Ninety Years* (Anchorage: Alaska Northwest, 1976), p. 35. In late July of 1898, the Northwest Mounted Police, who registered each traveler on the Overland route, estimated that 18,000 men and 630 women had already successfully made it across Dyea and Skagway trails.
2. Seattle *Post-Intelligencer,* February 5, 1895, p. 2; Juneau *Alaska News,* May 10, 1894.
3. Esther Lyons, "An American Girl's Trip to the Klondike," *Leslie's Weekly,* January 6, 1898, pp. 5, 7; Cad Wilson, Esther Lyons obituary, source unidentified, Alaska Historical Library, Wickersham Collection, PCA 227–1–195.
4. E. L. Robinson obituary, *New York Times,* October 28, 1938.
5. Ibid.; Lyons, *Leslie's Weekly,* January 6, 1898, pp. 5, 7; January 13, 1898, p. 21f; January 20, 1898, p. 37f; January 27, 1898, p. 53f; February 3, 1898, p 69f; February 10, 1898, p. 85f.
6. Wilson obituary, op. cit.
7. Frank Lynch, "Seattle Scene," Seattle *Post-Intelligencer,* April 10, 1951, p. 19.
8. W. Ogilvie, *Early Days on the Yukon* (London: Lane, 1913), p. 39, notes Mrs. Healy's presence at Dyea in 1887, for example.
9. Inga Sjolseth Kolloen, diary, Manuscripts Division, University of Washington Libraries, Seattle. Translated from Norwegian by Ruth Herberg, with additions by Dennis Andersen.
10. Ogilvie, op. cit.
11. Tappan Adney, *The Klondike Stampede of 1897–1898* (New York: Harper, 1899), p. 91. See Seattle *Times,* July 31, 1897, p. 1, for cost of typical outfit.
12. F. Hartshorn, Dyea manuscript, pp. 6–7, file 1–3, Manuscripts Division, University of Washington Libraries, Seattle. Compare with similar account in E. L. Martinsen, *Trail to North Star Gold* (Portland, Ore.: Metropolitan, 1969), p. 38. Another, somewhat confused, version of this story appears in Skagway *News,* February 11, 1898, p. 1.
13. B. A. Mulrooney Carbonneau in S. Franklin, "She Was the Richest Woman in the Klondike," Vancouver, B. C., *Sunday Sun, Weekend Magazine,* 12 no. 27 (July 7, 1962), pp. 28–29. It is not clear from the interview whether Belinda is describing an incident on Dyea or White Pass trail. But her first trip to the Klondike was across Dyea trail.
14. Kolloen, op. cit.
15. A. E. Berry, *The Bushes and the Berrys* (San Francisco: C. J. Bennett, 1941/1978), pp. 49–50.
16. Ibid., p. 50.

17. Ibid., p. 52.
18. Charles Meadows, Santa Cruz [Calif.] *Surf,* December 3, 1897, p. 1; reprinted from Kingman Arizona, *Our Mineral Wealth.*
19. Mae McKamish Meadows, letter, September 28, 1897, in Santa Cruz [Calif.] *Daily Sentinel,* October 24, 1897, p. 1.
20. Ibid.
21. Kolloen, op. cit.
22. Black, op. cit., p. 20.
23. Ibid., p. 28.
24. Ibid., pp. 28–29.
25. Kolloen, op. cit.
26. E. F. Larson, interview, September 17, 1979, pp. 30–31.
27. Berry, op. cit., p. 51.
28. Seattle *Post-Intelligencer,* April 10, 1898.
29. Adney, op. cit., p. 45.
30. E. L. Kelly, "A Woman's Trip to the Klondike," *Lippencott's Monthly,* November 1901, pp. 625–33.
31. Kelly, "A Woman Pioneer in the Klondike and Alaska," *Short Stories,* December 1912, pp. 76–84.
32. Kelly, "Woman's Trip," p. 626.
33. Ibid., p. 627.
34. Black, op. cit., p. 29.
35. Cambridge [Mass.] *Press,* April 16, 1898, p. 1f.
36. Ella Hall, "Account of a Trip to the Klondike Made in 1898." Manuscripts Division, Public Archives of Canada, Ottawa, Ontario. See also Lyon's account of similar frivolity with Native American packers in 1894, *Leslie's Weekly,* January 13, 1898, p. 23.
37. Black, op. cit., pp. 29–30.
38. Larson, interview, October 23, 1979, p. 2.
39. Ibid.
40. Kolloen, op. cit.
41. Georgia White, diary, June 22, 1898, p. 5, transcribed by Dorothy DeBoer, Yukon Archives, Whitehorse; original at Alaska State Museum, Juneau.
42. Ibid., June 27, 1898, pp. 7–8.
43. Larson, interview, September 17, 1979, p. 31.
44. Hartshorn, op. cit., p. 6, file 1–15, notebook 4. See also Juneau *Alaska Searchlight,* June 5, 1897, p. 8; Fairbanks *Daily Times,* November 10, 1907, p. 8; and Pioneer Women of Alaska Scrapbook, Clara Rust Collection, University of Alaska, Fairbanks, Archives.
45. Fairbanks *Daily Times,* November 10, 1907, p. 8.
46. Hartshorn, op. cit., pp. 57–58, file 1-15, notebook 7.

47. Ibid. Kate Ryne stayed for awhile in the Atlin District near Skagway. She had a small stake in a claim on Pine Creek, but eventually she used the little gold she got from it to pay her fare to the outside.

Chapter 5

1. J. Bernard Moore, *Skagway in Days Primeval* (New York: Vintage, 1968), pp. 15–16.
2. Ibid.
3. E. F. Larson, interview, September 17, 1979, pp. 3–4.
4. Ibid., pp. 4–6.
5. Ibid., pp. 6–7.
6. Ibid., p. 7.
7. Flora Shaw, Reuter's dispatch of May 6, 1898, London *Times,* June 6, 1898, p. 8.
8. Annie Hall Strong, "Impressions of Skagway," Skagway *News,* December 31, 1897, p. 9.
9. H. Clifford, *The Skagway Story* (Anchorage: Alaska Northwest, 1975), pp. 10, 13.
10. Larson, interview, September 17, 1979, pp. 10–11.
11. Larson, interview, October 23, 1979, p. 10.
12. Ibid.
13. See, e.g., Moore, op. cit., p. 21.
14. Larson, interview, September 17, 1979, p. 8.
15. Ibid., p. 19.
16. Ibid., pp. 19–20.
17. Larson, interview, September 24, 1979, p. 8.
18. Larson, interview, October 10, 1979, pp. 10–11.
19. Stroller [E. J.] White, *Stroller's Weekly* 20, no. 27 (May 28, 1927); reprinted as "Barbara," in *"Stroller" White: Tales of a Klondike Newsman,* ed. R. N. DeArmond (Vancouver, B.C.: Mitchell, 1969), pp. 19–24.
20. Ibid., p. 20.
21. Ibid., p. 21.
22. Ibid., p. 23.

Chapter 6

1. Tappan Adney, *The Klondike Stampede of 1897–1898* (New York: Harper, 1899), pp. 55–56.
2. Ibid., p. 61.

3. Ibid., p. 73.
4. Ibid., pp. 55–56.
5. Emma L. Kelly, "A Woman Pioneer in the Klondike and Alaska," *Short Stories,* December 1912, pp. 80–81.
6. E. M. Bell, *Flora Shaw* (London: Constable, 1947), p. 204.
7. Flora L. Shaw (Lugard), "Klondike," January 31, 1899, *Proceedings of the Royal Colonial Institute* 30, (1898/99), p. 112.
8. Ibid., pp. 113–14.
9. Ibid., pp. 112–13.
10. Bell, op. cit., p. 302.
11. Shaw, op. cit., p. 113.
12. E. A. Becker, "Little Girl in the Klondike Gold Fields," *Alaska Sportsman* 28, no. 11, (November 1962), p. 24. The tramcar which she remembers may have been around White Horse Rapids rather than on White Pass trail itself.
13. E. F. Larson, interview, September 24, 1979, p. 10; October 23, 1979, p. 9.
14. Larson, interview, September 17, 1979, pp. 20–22.
15. Becker, "A Klondike Woman's Diary," *Alaska,* June 1970, p. 39.
16. Ibid., pp. 39–40.
17. Larson, interview, September 17, 1979, p. 13.
18. Ibid., pp. 15–16.
19. Ibid., p. 15.
20. Ibid., pp. 24–25.
21. Ibid., pp. 28–29.
22. Ibid., pp. 23–24.
23. Ibid., p. 7.
24. Adney, op. cit., pp. 62–63.
25. Larson, interview, September 17, 1979, pp. 22–23. An article in the September 16, 1898, Skagway *News,* p. 2, reported that Skaguay Belle was by that time in Dawson and doing very well running a laundry.
26. Becker, "Klondike Woman's Diary," pp. 40–41.
27. Larson, interview, September 17, 1979, pp. 25–26.
28. Ibid., pp. 26–27.
29. B. A. Mulrooney Carbonneau, in S. Franklin, "She Was the Richest Woman in the Klondike," Vancouver, B.C., *Sunday Sun, Weekend Magazine,* 12, no. 27 (July 7, 1962), p. 28. Belinda apparently was not one to hold a grudge. An article in the September 16, 1898, Skagway *News,* p. 2, reports that she engaged Brooks at that time to haul in more supplies for her. But in September the contract for hauling included the clause, "The goods must all be at Bennett within three days, or no pay, see?" In this article, Belinda is mistakenly identified as "Miss Maloney."

30. Larson, interview, September 17, 1970, p. 8.
31. Ibid., p. 9.
32. Larson, interview, September 17, 1979, pp. 20–21.
33. Skagway *Daily Alaskan*, December 8, 1898, p. 1.
34. Larson, interview, September 24, 1979, p. 16.
35. Skagway *Daily Alaskan*, December 8, 1898, p. 1.
36. Larson, interview, September 24, 1979, p. 16.
37. Ibid.
38. Ibid., pp. 17, 19.
39. Larson, interview, September 17, 1979, pp. 9–10.
40. A. C. Stevens, *The Cyclopedia of Fraternities* (New York: E. B. Treat and Co., 1907), pp. 154–56; and A. J. Schmidt, *Fraternal Organizations* (Westport, Conn.: Greenwood Press, 1980), p. 213.

Chapter 7

1. Mae McKamish Meadows, letter, October 1, 1897, Santa Cruz [Calif.] *Daily Sentinel*, October 24, 1897, p. 1.
2. Santa Cruz [Calif.] *Daily Sentinel*, May 11, 1898, p. 1.
3. *Klondike: The Chicago Record's Book for Gold Seekers* (Chicago: Chicago Record Co., 1897), p. 272. The woman is not identified by name, but she was writing to her brother John C. Hessian in Duluth, Minn., while she was with her husband, Pat, a hardware merchant at Fort Cudahy (at Forty Mile River). This then is almost certainly Mary Ellen Galvin.
4. Dawson *Klondike Nugget*, June 28, 1898, p. 3.
5. Alice Edna Berry, *The Bushes and the Berrys* (San Francisco: C. J. Bennett, 1941/1978), pp. 52–53.
6. Ibid., p. 54.
7. Ibid., p. 55.
8. F. L. Shaw, Reuter's dispatches of May 14, 27, 1898, in the London *Times*, June 11, 1898, p. 10, and June 23, 1898, p. 5.
9. I. S. Kolloen, diary, Manuscripts Division, University of Washington Libraries, Seattle. Translated from Norwegian by Ruth Herberg, with additions by Dennis Andersen.
10. Martha Louise Black, *My Ninety Years* (Anchorage: Alaska Northwest, 1976), pp. 33–34. If Martha did meet Flora Shaw at Bennett, then the date Martha gives for her crossing of Chilkoot is incorrect. She states (p. 25) that she left Dyea July 12, 1898, which would put her at Bennett about July 14. But Flora Shaw had left Bennett soon after June 1 and arrived in Dawson on June 23. The London *Times* article of

August 10, 1898, says she arrived *July* 23, but it is obvious from Flora's July 1 and 13 dispatches in the *Times* of August 2 and 6 that *June* was intended. Martha may have reconstructed her erroneous dates from this *Times* misprint. This interpretation seems even more plausible because there was not enough time between July 12 and August 5, when Martha arrived in Dawson, to account for all the trail events she describes. And, since Flora Shaw did not leave Dawson until August 11, it could not have been that she might have met Martha at Bennett on her return trip.

11. Shaw, Reuter's dispatch of May 27, 1898, in London *Times*, June 23, 1898, p. 5.
12. F. Gillis and P. McKeever, "The Lady Went North in '98," *Alaska Sportsman*, February 1948, p. 13.
13. Dawson *Klondike News*, April 1, 1898, p. 4.
14. Kolloen, op. cit.
15. Miles Canyon and White Horse Rapids have now been submerged by a dam at White Horse.
16. *The Official Guide to the Klondyke Country and the Gold Fields of Alaska* (Chicago: Conkey, 1897), pp. 147–48.
17. M. F. McKeown, *The Trail Led North* (New York: Macmillan, 1948), pp. 158–59. Compare L. Day's account of the Lippy mishap in *The Tragedy of the Klondike* (New York: L. Day, 1906), pp. 51–52.
18. E. L. Kelly, "A Woman's Trip to the Klondike," *Lippencott's Monthly*, November 1901, pp. 625–33.
19. Ibid., p. 82.
20. Ibid., p. 630.
21. Ibid., p. 83. Emma made it to Dawson on November 1, 1897 and stayed through the following summer. When she returned to the outside in October of 1898, she was the owner of valuable mining claims. Seattle *Post-Intelligencer*, October 10, 1898, p. 6.
22. Kate Rockwell Matson, "I Was the Queen of the Klondike," *Alaska Sportsman*, August 1944, pp. 10–11, 28–32; also E. Lucia, *Klondike Kate* (Sausalito, Calif.: Comstock, 1962), pp. 60–62.
23. Black, op. cit., p. 35.
24. Luella Day, *The Tragedy of the Klondike* (New York: Luella Day, 1906), p. 53.
25. Kolloen, op. cit.
26. Gillis and McKeever, op. cit., p. 30.
27. Dawson *Klondike Nugget*, August 6, 1898, p. 3.
28. Georgia White, diary, February 21–October 8, 1898, Yukon Archives, ms. box II, file 11, June 29, 1898, pp. 8–9.
29. A. S. Stewart, Pioneer Women of Alaska Scrapbook, Clara Rust Collection, University of Alaska, Fairbanks, Archives.

30. Ibid.
31. Carolyn Niethammer, "The Lure of Gold," in *The Women Who Made the West,* Western Writers of America (Garden City, NY: Doubleday, 1980), pp. 71–72; and D. N. Bentz, "Frontier Angel," *The West,* July 1972, pp. 6–8, 60–61, citing Victoria *Daily British Colonist.*
32. Ivan C. Lake, "Nellie Cashman Blazed the Frontier Trails," *Real West,* November 1965, pp. 58–66.
33. Bernice Cosulich, Tuscon *Star,* January 18, 1924.
34. See, e.g., John Clum, "Nellie Cashman," *Arizona Historical Review,* January 31, 1931, pp. 9–34; Lake, op. cit.; Harriet Rochlin, "The Amazing Adventures of a Good Woman," *Journal of the American West,* April 1973, pp. 281–95; Niethammer, op. cit., pp. 71–87.
35. Tucson *Daily Star,* November 23, 1889.
36. Phoenix *Daily Herald,* May 30, 1895.
37. For example, H. Rochlin, op. cit.
38. Nellie Cashman's birthday is variously reported to be between 1844 and 1851. See, for example, John L. Gilchriese, Tucson *Citizen,* April 5, 1965; Lake, op. cit., pp. 19, 42–45; and Clum, op. cit., p. 10.
39. Florence [Ariz.] *Tribune,* November 20, 1897.
40. See, e.g., Niethammer, op. cit.; Lake, op. cit.; and Victoria *British Daily Colonist,* February 15, 1898.
41. Proceedings of *Alaskan Boundary Tribunal* (Washington, D. C.: Government Printing Office, 1904).
42. In fact, duty was revoked by the U.S. government for Dyea and Skagway by late July and early August of 1897, though it took somewhat longer for the information to reach officials at these ports.
43. Skagway *News,* August 26, 1898. The story was probably written by Stroller (E. J.) White.
44. Lucille Hunter obituary, Whitehorse *Star,* June 12, 1972, p. 5; interview with Victoria Faulkner, Whitehorse, August 30, 1980.

Chapter 8

1. Georgia White, diary, July 2, 1898, p. 10, ms. box II, file 11, Yukon Archives, Whitehorse.
2. F. D. Gillis and P. McKeever, "The Lady Went North in '98," *Alaska Sportsman,* February 1948, p. 30.
3. Pierre Berton, *The Klondike Fever* (New York: Knopf, 1958), p. 92.
4. Ibid., p. 295.
5. Seattle *Post-Intelligencer,* September 2, 1898, p. 6.
6. Berton, op. cit., p. 400.

7. Ibid.
8. *Klondike Nugget,* October 25, 1898, p. 3; unidentified newspaper article, University of Washington Alaska Biography, George W. Carmack, N979.8; and Seattle *Post-Intelligencer,* September 1, 1898, p. 5.
9. *Klondike Nugget,* July 26, 1899, p. 3.
10. Ibid., October 14, 1900, p. 2; Frank Lynch, "Seattle Scene," Seattle *Post-Intelligencer,* April 11, 1951.
11. *Pathfinder,* August 1920, p. 22.
12. Berton, op. cit., p. 59.
13. Ibid., p. 418–19.
14. San Francisco *Chronicle,* July 21, 1897, p. 2.
15. Some stories say Belinda arrived as early as the summer of 1896, but this seems unlikely since the discovery of the Klondike gold was not until August of 1896. She was on the steamer *Mexico,* headed for Dyea, when it arrived in Juneau, March 29, 1897 (*Alaska Searchlight*), April 3, 1897, p. 8). The photo of her on Chilkoot trail was taken in mid-April of 1897. This would fit with mining records (*Original Locators,* Henderson Creek, vol. 133, p. 295, Yukon Archives, Whitehorse) showing that she located a claim at 22 below on Henderson Creek on June 14, 1897. According to an early *Klondike News* story, April 1, 1898, this staking was done on her way in.
16. Stephen Franklin, "She Was the Richest Woman in the Klondike," Vancouver, B. C., *Sunday Sun,* (July 7, 1962), *Weekend Magazine* 12, no 27, pp. 22–23, 28–29.
17. N. Bolotin, *Klondike Lost* (Anchorage: Alaska Northwest, 1980). Also, *Index of Original Locators,* Placer Claims, vol. 1, p. 295; *Mining Recorder Record,* Bonanza Creek, vol. 1, pp. 17–18; Yukon Archives, Whitehorse.
18. *Klondike News,* April 1, 1898, pp. 4, 30.
19. *Klondike Nugget,* July 27, 1898.
20. M. E. Hitchcock, *Two Women in the Klondike* (New York: Putnam, 1899), pp. 100–102.
21. Seattle *Post-Intelligencer,* December 7, 1908.
22. Ibid.
23. Betty Mulrooney, interview by M. J. Mayer, October 20, 1979.
24. *Klondike Nugget,* May 17, 1899; also Berton, op. cit., p. 425.
25. W. V. Mackay, "The Saga of Belinda," Seattle, *Sunday Times, Charmed Land Magazine,* August 12, 1962; National Register of Historic Places Inventory, Nomination Form, Carbonneau Castle, Yakima, Wash.
26. *Mining Recorder Record,* Bonanza Creek Claims, vol. 3; *Mining Recorder Record,* Creek Claims, Eldorado, vol. 22; Yukon Archives, Whitehorse.

27. Dawson *News*, March 3, 1905.
28. Ibid., September 22, 1904.
29. Ibid., October 12, 1904; ibid., February 15, 1905.
30. John Clark, letter, December 15, 1922, University of Alaska, Archives and Manuscript Collection, Fairbanks.
31. Record Group 509, United States District Court, Fourth Division, Series 417, Journal (OS 13), December 10, 1906.
32. Dale and Sarah Kroll, interview by M. J. Mayer, October 20, 1979.
33. Death certificate, Washington State Department of Health, No. 19864.
34. *Klondike News*, April 1, 1898, p. 31; also Seattle *Post-Intelligencer*, April 19, 1898.
35. *Klondike Nugget*, December 30, 1899, p. 3.
36. Ibid., September 27, 1899, p. 1.
37. *Daily Klondike Nugget*, October 18, 1901, p. 1.
38. Jean King, personal communication, 1988.
39. Santa Monica *Evening Outlook*, December 22, 1960, p. 25; Jean King, personal communication, 1988.
40. *Klondike News*, April 1, 1898, p. 21; *Klondike: The Chicago Record's Book for Gold Seekers* (Chicago: Chicago Record Co., 1897), p. 272.
41. *Klondike News*, April 1, 1898, p. 21.
42. Berton, op. cit., p. 418.
43. Alaska Historical Library, Wickersham Collection, PCA 277–11–39.
44. *Alaska-Yukon Directory and Gazeteer*, 1902; U.S. Census, Fairbanks, 1910; Dawson *Daily News*, September 24, 1904, p. 1.
45. *Mining Recorder*, Original Locators, Folio 6320, Yukon Archives.
46. Victoria Faulkner, personal communication, August 30, 1980; Whitehorse *Star*, June 12, 1972, p. 5.
47. Erling Kolloen, personal communication, July 5, 1983.
48. Gillis and McKeever, op. cit., p. 30.
49. Ibid., p. 31.
50. Ibid., pp. 31–32.
51. *Klondike Nugget*, June 28, 1898, p. 1; John P. Clum, "Nellie Cashman," *Arizona Historical Review*, January 1931, p. 27.
52. *Yukon Territory Central Registry Records*, RG 91, 73/37, letter to E. C. Senkler, gold commissioner, from F. J. McDougal, solicitor, January 31, 1900, Yukon Archives, Whitehorse.
53. Ibid., letter to secretary, Department of the Interior, from E. C. Senkler, gold commissioner, July 8, 1901.
54. Ibid., deposition of Nellie Cashman to E. C. Senkler, gold commissioner, July 8, 1901.
55. Ibid., letter to secretary, Department of the Interior, from E. C. Senkler, gold commissioner, October 17, 1900.

56. Ibid., award dated November 30, 1901.
57. Clum, op. cit.
58. I. C. Lake, "Irish Nellie, Angel of the Cassiar," *Alaska Sportsman,* October 1963, pp. 19, 42–45.
59. Clum, op. cit., p. 30.
60. Ibid., p. 31.
61. Lake, "Nellie Cashman Blazed the Frontier Trails," *Real West* 8 (November 1965), pp. 58–66.
62. Helen Berg, "The Doll of Dawson," *Alaska Sportsman* 10 (February 1944), pp. 8–9, 25–31. After Mae Field's husband, Arthur, left her (probably in about 1904), she worked at the Floradora as a dance hall "girl." At that time she was given the name, "the Doll of Dawson."
63. *Yukon World,* June 2, 1904.
64. Fairbanks *Daily Times,* July 22, 1908.
65. Bernice Cosulich, Tucson *Arizona Star,* January 11, 1925.
66. Ibid.
67. Skagway *Daily Alaskan,* June 6, 1919.
68. *Sunset Magazine,* May 1921, p. 48.
69. W. C. Fonda, scrapbook, article dated January 10, 1924, Seward, Alaska. University of Washington, Northwest Collection.
70. J. S. Reiter and Time-Life Book Editors, *The Women* (Alexandria, Va.: Time-Life, 1978), p. 170.
71. Harriet Rochlin, "The Amazing Adventures of a Good Woman," *Journal of the American West,* April 1973, p. 295.
72. M. L. Black, *My Ninety Years* (Anchorage: Alaska Northwest, 1976), p. 156.
73. Ibid., p. 59.
74. Ibid., p. 158.
75. Ibid., p. 111.
76. Ibid.
77. Berton, op. cit., p. 384.
78. E. L. Robinson obituary, *New York Times,* October 28, 1938; E. Lyons, *Leslie's Weekly,* January 6, 1898, p. 5f; January 13, 1898, p. 21f; January 20, 1898, p. 37f; January 27, 1898, p. 53f; February 3, 1898, p. 69f; February 10, 1898, p. 85f. Cad Wilson (Esther Lyons) obituary, source unidentified, Alaska Historical Library, Wickersham Collection, PCA 227–1–195.
79. E. L. Robinson obituary, op. cit.
80. *Klondike Nugget,* October 29, 1898, p. 2.
81. Ibid., October 21, 1899.
82. Ibid., April 19, 1899, p. 4.
83. Ibid.
84. E. L. Robinson obituary, op. cit.

85. E. A. Becker, "Little Girl in the Klondike Gold Fields," *Alaska Sportsman,* November 1962, pp. 22–24, 34–38.
86. Ibid.
87. Ibid.
88. Washington State Death Index, 171, June 5, 1970.
89. *Klondike Nugget,* August 17, 1898, p. 4.
90. Ibid., October 1, 1898, p. 3.
91. Luella Day, *The Tragedy of the Klondike.* (New York: Luella Day, 1906), pp. 125–31, 143.
92. Ibid., pp. 176ff.
93. White, op. cit., July 8–September 6, 1898, pp. 12–16.
94. Dorothy DeBoer, personal communication, September 1983.
95. *Klondike Nugget,* August 5, 1899, p. 4.
96. Ibid., August 23, 1899, p. 3.
97. E. C. Romig, *A Pioneer Woman in Alaska* (Caldwell, Idaho: Caxton, 1948).
98. Seattle *Post-Intelligencer,* May 8, 1957.

REFERENCES

Photograph Sources

Photographs and maps are reproduced with permission of the following (abbreviations used in the figure legends and herein are in parenthesis):

Alaska Historical Library (A.H.L.) Alaska State Library, Juneau
Anchorage Museum of History and Art (A.M.H.A.)
Arizona Historical Society (A.H.S.), Tucson, C. J. Peter Bennett
Dorothy DeBoer
Dedman's Photo, Skagway
Kermit Edmonds
Jean King
Edith Feero Larson
London *Times*
Post-Intelligencer (P.I.), Seattle
Provincial Archives of Alberta (P.A.A.), Edmonton
Seattle *Times*
United States Park Service
University of Alaska (U.A.F.), Fairbanks
University of Washington (U.W.), Seattle, Photographs and Graphics Collection
Gladys Wilfong and Marjorie Lambert
Yukon Archives (Y. A.), Whitehorse

Photographers

Adams, Edward and Larkin, George
Banks, H. D.
Barley, H. D.
Blankenberg, J. M.
Brooks, A. H.
Cantwell, George G.
Child
Clum, John P.
Curtis, Ashel
Dow, Dr.
Ellingson, E. O.
Forbes, Stuart
Goetzman, H. J.
Hegg, Eric A.
Hitchcock, Mary E.
Kelly, M. F.
Kilburn, Ben W.
LaRoche, Frank

Larss, P. E. and Duclos, J. E. N.
Martin, A. A.
Mathers, C. W.
Metcalf, Charles F.
Ogilvie, William
Romig, E. C.
Rose, N. H.
Sarvant, Henry
Waite, Alvin H.
Wilse, Anders B.
Wilson, Veazie
Winter, Lloyd V. and Pond, E. P.
Wolfe

Published Sources

Adney, Tappan. *The Klondike Stampede of 1897–1898.* New York: Harper, 1899.

Alaska-Yukon Directory and Gazeteer, 1902.

Becker, E. A. "Little Girl in the Klondike Gold Fields." *Alaska Sportsman* 28, no. 11, (November 1962), pp. 22–24, 34–38.

Becker, E. A. "Discoverer of the Klondike." Seattle *Post-Intelligencer,* "Pictorial Review" (August 18, 1963), pp. 10–12.

Becker, E. A. "A Klondike Woman's Diary." *Alaska* (June 1970), pp. 18–19, 39–41.

Bell, E. M. *Flora Shaw,* London: Constable, 1947.

Bentz, D. N. "Frontier Angel," *The West* (July 1972), pp. 6–8, 60–61.

Berg, H. "The Doll of Dawson," *Alaska Sportsman* 10 (February 1944), pp. 8–9, 25–31.

Berry, A. E. *The Bushes and the Berrys.* San Francisco: C. J. Bennett, 1941/1978.

Berton, Pierre. *The Klondike Fever.* New York: Knopf, 1958.

Black, Martha L. *My Ninety Years.* Anchorage: Alaska Northwest, 1976.

Bolotin, N. *Klondike Lost.* Anchorage: Alaska Northwest, 1980.

Carmack, George W. *My Experience in the Yukon.* Privately published, 1933.

Clifford, H. *The Skagway Story.* Anchorage: Alaska Northwest, 1975.

Clum, John P. "Nellie Cashman." *Arizona Historical Review* 3 (1931), pp. 9–34.

Cosulich, Bernice. *Tucson Star,* January 18, 1924.

Cosulich, Bernice. *Arizona Star,* Tucson, January 11, 1925.

Crane, Alice Rollins. *Smiles and Tears.* New York: Doxey's, 1901.

Day, Luella. *The Tragedy of the Klondike.* New York: Luella Day, 1906.

DeArmond, R. N. *'Stroller' White: Tales of a Klondike Newsman.* Vancouver, B. C.: Mitchell, 1969.

Dyer, E. J. *The Routes and Mineral Resources of Western Canada.* London: Philip, 1898.

Emmons, S. F. *Map of Alaska.* Washington, D.C.: U.S. Geological Survey, 1898.

Franklin, Stephen. "She Was the Richest Woman in the Klondike." Vancouver, B.C., *Sunday Sun, Weekend Magazine,* 12 no. 27 (July 7, 1962), pp. 22–23, 28–29.

Galpin, William. "Who Is to Blame?" in *Smiles and Tears from the Klondyke,* edited by Alice Rollins Crane. New York: Doxey's, 1901.

Gillis, Francis D. and P. McKeever. "The Lady Went North in '98," *Alaska Sportsman* (February 1948), pp. 12–13, 30–32.

Hitchcock, Mary E. *Two Women in the Klondike.* New York: Putnam, 1899.

Jones, H. W. *Women Who Braved the Far North.* Vol 1. San Diego: Grossmont Press, 1976.

Kelly, E. L. "A Woman's Trip to the Klondike." *Lippencott's Monthly* (November 1901), pp. 625–33.

Kelly, E. L. "A Woman Pioneer in the Klondike and Alaska," *Short Stories* (December 1912), pp. 76–84.

Klondike: The Chicago Record's Book for Gold Seekers, Chicago: Chicago Record Co., 1897.

The Klondyker. Vol. 1. Seattle: Alaska Intelligence Bureau, November 1, 1897.

Ladue, Joseph. *Klondyke Facts.* New York: American Technical Book Co., 1897.

Lake, Ivan C. "Irish Nellie, Angel of the Cassiar." *Alaska Sportsman* (October 1963), pp. 19, 42–45.

Lake, Ivan C. "Nellie Cashman Blazed the Frontier Trails." *Real West* (November 1965), pp. 58–66.

LaRoche, F. *En Route to the Klondyke.* Chicago: Conkey, 1898.

Little, J. R. "Squaw Kate." *Alaska Life* (March 1943), pp. 17–18.

Lucia, E. *Klondike Kate.* Saulsalito, Calif.: Comstock, 1962.

Lynch, Frank. "Seattle Scene." Seattle *Post-Intelligencer,* April 10, 1951, p. 19.

Lynch, Frank. "Seattle Scene." Seattle *Post-Intelligencer,* April 11, 1951.

Lynch,J. *Three Years in the Klondike.* Chicago: Lakeside, Donnelley, 1967.

Lyons, Esther. "An American Girl's Trip to the Klondike," *Leslie's Weekly* January 6, 1898, pp. 5, 7; January 13, 1898, p. 21, 23; January 20, 1898, p. 37, 39; January 27, 1898, p. 53, 55; February 3, 1898, p. 69, 71; February 10, 1898, p. 85, 87.

MacGregor, J. G. *The Klondike Rush through Edmonton, 1897–1898.* Toronto: McClelland and Stewart, 1970.

MacKay, Wallace V. "The Saga of Belinda." Seattle *Times, Charmed Land Magazine,* August 12, 1962, pp. 4–6.

Martinsen, E. L. *Trail to North Star Gold.* Portland, Ore.: Metropolitan, 1969.

Mathers, C. W. *A Souvenir from the North.* Edmonton, Alb.: 1901.

Matson, Kate R. "I Was the Queen of the Klondike." *Alaska Sportsman* (August 1944) pp. 10–11, 28–32.

McKeown, M. F. *The Trail Led North.* New York: Macmillan, 1948.

Moore, J. Bernard. *Skagway in Days Primeval.* New York: Vintage, 1968.

Niethammer, C. "The Lure of Gold." In *The Women Who Made the West,* edited by Western Writers of America. Garden City, N.Y.: Doubleday, 1980, pp. 71–87.

Nin, Anaïs. *The Diary of Anaïs Nin, 1931–1934.* Edited by Gunther Stuhlmann, New York: Swallow and Harcourt, Brace and World, 1966.

Official Guide to the Klondyke Country and the Gold Fields of Alaska. Chicago: Conkey, 1897.

Official Map Guide, Seattle to Dawson. Seattle: Humes, Lysons, and Sallee, 1897.

Ogilvie, William. *Information Respecting the Yukon District.* Ottawa: Government Printing Office, 1897.

Ogilvie, William. *The Klondike Official Guide.* Toronto: Hunter Rose, 1898.

Ogilvie, William. *Early Days on the Yukon.* Ottawa: Thorburn Abbott, 1913.

Orth, Donald J. *Dictionary of Alaska Place Names.* U.S. Geological Survey Professional Paper 567. Washington, D.C.: Government Printing Office, 1967.

Page, Elizabeth. *Wild Horses and Gold: From Wyoming to the Yukon.* New York: Farrar and Rinehart, 1932.

The Pathfinder. August 1920, p. 22.

Proceedings of Alaskan Boundary Tribunal, Washington, D.C.: Government Printing Office, 1904.

Reiter, J. S. and Time-Life Book Editors. *The Women.* Alexandria, Va.: Time-Life, 1978.

Romig, Emily Craig. *A Pioneer Woman in Alaska.* Caldwell, Idaho: Caxton, 1948.

Rochlin, Harriet. "The Amazing Adventures of a Good Woman." *Journal of the American West* (April 1973), pp. 281–95.

Schmidt, A. J. *Fraternal Organizations.* Westport, Conn.: Greenwood, 1980.

Shaw, F. L. "Klondike," January 31, 1899. In *Proceedings of the Royal Colonial Institute* 30, (1898/99), pp. 109–35.

Stevens, A. C., ed. *The Cyclopedia of Fraternities.* New York: Treat, 1907.

Strong, Annie Hall. "Impressions of Skagway." Skagway *News,* December 31, 1897, p. 9.

To the Klondike and Gold Fields of the Yukon. Canadian Pacific Railway, 1897.

White, Stroller (E. J.). *Stroller's Weekly* 20, no 27. (May 28, 1927). Reprinted as "Barbara." In *"Stroller" White: Tales of a Klondike Newsman,* edited by R. N. DeArmond, Vancouver, Alb.: Mitchell, 1969, pp. 19–24.

The Yukon District of Canada. High Commissioner for Canada, London 1897.

Yukon Gold Fields Map; Mining Regulations, and Extracts of Mr. Ogilvie's Reports. Victoria, B.C.: Colonist, 1897.

Unpublished Sources

Burr, A. Regina Alles. Diary. A. H. F. A. M.

Clark, J. Letter, December 15, 1922. University of Alaska, Archives and Manuscript Collection, Fairbanks.

Day, Evelyn. Interview by author, July 13, 1982, Victoria, B.C.

Faulkner, Victoria. Interview by author August 30, 1980. Whitehorse, Y. T.

Hall, Ella. "Life in the Klondike. . . ." Manuscript, Public Archives of Canada, Ottawa.

Hartman, J. D. Letter to E. A. Becker, September 29, 1942, University of Washington Alaska Biography, N979.819.

Hartshorn, Florence M. Papers, 1909–1934. U. W.

Kolloen, Inga S. Diary. U. W.

Kroll, Dale and Sarah. Interview by author October 20, 1979, Yakima, Wash.

Larson, Edith F. Interviews by author September 17, and 24, October 1 and 23, 1979, Tacoma, Wash.

Mulrooney, Betty. Interview by author October 20, 1979, Spokane, Wash.

Rickert, Stacia T. Barnes. Pioneer Women of Alaska Scrapbook, Clara Rust Collection, U.A.F. Archives.

Stewart, A. S. Pioneer Women of Alaska Scrapbook, Clara Rust Collection, U.A.F. Archives.

White, Georgia. Diary. Transcribed by D. DeBoer. A. H. L. and Y. A.

Related Literature

Alberts, Laurie. "Petticoats and Pickaxes." *The Alaska Journal* 7, no. 3 (Summer 1977), pp. 148–59.

Banks, Della Murray. "Woman on the Dalton Trail." *Alaska Sportsman* (January 1945), pp. 10–11, 25–34.

Banks, Della Murray. "Rainbow's End." *Alaska Sportsman* (February 1945), pp. 14–15, 21–27.

King, Jean. *Arizona Charlie Meadows: Adventures of a Rainbow Chaser.* In press.

Kneass, Amelie. "The Flogging at Sheep Camp." *Alaska Sportsman* (October 1964), pp. 18–19, 57–58.

Lung, Velma D. "The Expected Guest." *Alaska Sportsman* (April 1950), pp. 10–13, 28–30.

Mark-Anthony, Bev. "Gold Rush: Women of the Klondike." *Alaska Woman* (February/March 1978), pp. 20, 87–90.

McKay, Kathy. "What about the Women of the Klondike Gold Rush?" Skaguay *Alaskan* 11 (1988).

Rosen, Ruth. *The Lost Sisterhood.* Baltimore: John Hopkins University, 1982.

Sullivan, May Kellogg. *A Woman Who Went to Alaska.* Boston: James Earle, 1902.

INDEX

(Bold numbers refer to Figures.)

A

Actresses, see Entertainers
Adney, Tappan, 28, 96, 131, 133, 134–35, 150
Alaskan Highway, 29, 230, **2–1**
Alaskan Territory, 17, 194
Alice, **3–4**
Al-ki, 23–24, 92, 113, **1–12**
Amery, C. W., 164
Anchorage, Alaska, 235
Andersen, 73, 105
Anderson family, 25–26, 143
  Clay, 26, 143, **8–27**
  Dewey, 26, **8–27**
  Emma, 25–26, 143, 233–234
  Ethel, see Becker, E. A.
  Peter B., 25, 233
Angel of the Cassiar, see Cashman, N.
Anglo French Klondyke Syndicate, 211
Animatoscope, see Magic lantern
Anson, 83
Arizona Charley, see Meadows, C.
Arctic Circle, 48, 224
Athabasca Landing, Alberta, 35, 51
Athabascan tribes, 185–186
  see also Tutchone, Southern
Athapascan, see Athabascan
Atlin, British Columbia, 164, 195
Australia, 235
Avalanche, 91–93, **4–26**

B

Babies, 24, 26, 27, 35, 36, 48, 51, 76, 108, 109, 143, 184, 205, 217, 221, 227, 239, **4–9, 4–25**
Baily, 190
Barbara, 128–29
Barnes, George W., 28
  Stacia T., see Rickert, S. T. B.
Bars, 69, 82, 84, 115, 119, 125, 142, 172, 210, **5–10**
Baudry, 85
Becker, Ethel Anderson, 4, 13–16, 25–26, 143, 233–34, **8–27**
Bench claims, 200–01
Bennett, British Columbia, 106, 138, 163, 167, 171–76, **6–26, 7–1** through **7–4**
Bering Sea, 54, 56
Berry family, 205–07, **8–8, 8–9**
  Alice Edna Bush, see Tot B.
  Clarence J., 3, 6, 81–82, 93, 170–71, 205–07, **8–8, 8–9**
  Ethel Bush, 3, 6–7, 81–82, 93, 170–71, 201, 205–07, **8–8, 8–9**
  Henry, 206, **8–9**
  Poie, see William J.
  Tot Bush, 6–7, 81–82, 93, 99, 170–71, 206–07, **8–8, 8–9**
  William J., 171
Berry Oil Company, 207
Bert, see Voorhees, B.
Berton, Pierre, 204
Bicycle, 101
Biegler, Gus, **4–12**
Big Stone, Chief, 36
Black, George, 227–31, **8–23, 8–24, 8–25**
  Martha Munger Purdy, 2, 89–91, 95, 98, 102–04, 174, 184, 227–30, 245–46, **4–25, 8–23, 8–24, 8–25**

Bledsoe's feed store, 143
Blockade, 131, 155, 162–63
Bly, Nellie, **6–7**
Boat building, 44, 49, 106, 167, 172, 176, 190, **7–1, 7–2, 7–6**
Bodin, Ida, 78, 105
Bonanza creek, 18–19, 201, 205, 206, 207, 211, 217, 220, 221, 222, **1–7, 1–8, 8–10, 8–17**
Bonanza King, 189–90
Border, Canada-United States, 97–98, 162–63, 194, **4–23, 4–28, 6–24, 6–25**
Bower, Bert, **4–12**
Brady, Frank, 93–95
  Marie Isharov, 93–95
Braund family, 49
  Mr., 47–48
  Mrs., 47–48, 238–39
Britton, Mr., 186, 198
Brooks, Joe, 156–57, **6–19**
Burkhard House, **5–5**
Burns, Archie, 95
Bush, Alice Edna, see Berry, Tot B.
  Ethel, see Berry, E. B.
  Tot, see Berry, Tot B.
The Bushes and the Berrys, 206

C

Cabins, 48, 48–49, 119–20, 204, 205, 207, **1–8, 5–9, 6–13, 8–8, 8–10, 8–17**
Calgary, Alberta, 35
Campbell, Rev. Dr., **5–12**
Camp tending, 32, 34, 41, 45, 49, 81–82, 170–71
Canada, S. S., **8–24**
Can-Can Restaurant, 221
Canyon, Chilkoot trail, 78–81, **4–2, 4–5, 4–13, 4–14, 4–15**
Caracas, Jonny, 92–93
Carbonneau, Belinda Mulrooney, 76–78, 156–57, 179, 201, 207–13, 241, 244, 248, **4–12, 6–18, 8–10, 8–11, 8–12, 8–13**
  Castle, 213, **8–13**
  Charles E., 211–13, **8–12**
Carcross, British Columbia, 163, 176, 204, **8–6**
Card, Harry W., 38
Card, Ella, 108–09, 201, 217–18, **4–9, 8–18**
  Fred, 108–09, 217–18, **4–9**
  Freddy, 218
Caribou Crossing, see Carcross, B. C.
Carmack, George W., 17–19, 68, 201–204, **1–8, 8–5**
  Graphie Gracie, 201–03, **1–8, 8–5**
  Kate Mason, 17–18, 68, 201–04, **1–8, 8–5, 8–6**
  Marguerite Laimee, 203–04
Carter, Mr., 38
Cashman, Nellie, 190–94, 221–27, 247, **7–16, 8–21, 8–22**
Cassiar District, 190, **7–15**
Cassiar Restaurant, 221
Catalpa Knob, Kansas, 174, 227
Cecil Hotel, 218
Cecil Restaurant, 218
Charlton, Ed, 41, 43, 44–45
Chase, 63
Cheechakos, 200
Cheever, C. G., 99–101
  Lizzie M., 2, 99–101
Chicago party, see Craig party
Children, 4, 24–26, 26, 27, 36, 69, 76, 105, 108–09, 111, 126–27, 141, 143, 148, 189, 193, **4–9, 4–14, 5–12, 5–14, 8–27**
  see also Babies
Chilkoot Pass, 68, 69, 86–98, 101, 131, 169, **4–2, 4–4, 4–5, 4–20** through **4–28**
Chilkoot trail, 4, 9, 60, 64–110, 131–32, 194, 231, **I–1, I–2, I–3, 4–1** through **4–39**
Christie, James, 6
Christmas, 124–26, 256, **5–13**
Churches, 121–22, 125, 194, **5–11, 5–12, 5–13**
City of Seattle, 113
Clantons, 193
Clark, Ella Card, see Card, E.
Clothing, 29, 31, 34, 36, 41, 48, 50, 55, 68, 85, 89–91, 95, 115, 144–45, 174, 184, 189, 227, **4–35**
Clum, John P., 221, 223
Cody, Buffalo Bill, 82, **8–16**
Columbia, 61–63

Conning, 120, 132, 154–56, **6–17**
Cook Inlet, 28
Cooking, 8, 25, 32, 35, 41, 43, 45, 47, 48, 72, 73, 82, 102, 145–46, 149, 170, 176, 179, 206, 219, 235
Corduroy road, 131–32
Correspondents, 4, 28, 96, 131, 133, 134–35, 138–42, 150, 172, 182, 230, **6–7**
  see also individual names
Courtesy, 36, 55, 91, 140–41, 226, 232
Cousins, 93, 143–45, **5–9**
Craig, Mr., 189
Craig party, 34–51
  A. C., 34–51, 235
  Emily, see Romig, E. C.
"Cremation of Sam McGee", 185
Crime, 120–21, 132, 154, 184, 211, 234
Cursing, see Profanity
Customs, 97–98, 194, 247, **4–23, 4–28**

D

Danaher, Captain, 63
Dance hall women, 81, 141, 232, 250
Dawson, Yukon, 8–9, 65, 163, 199–201, 209–11, 214, 217–18, 220–21, 227–33, 234, 235, **3–7, 6–18, 8–4, 8–20, 8–21, 8–26**
Day, Luella, Dr., see McConnell, L. D., Dr.
Dead Horse trail, 135–38, **6–9, 6–10**
  see also White Pass trail
Death, 27, 34, 36, 48, 51, 76, 91–93, 108, 109, 110, 164–65, 171, 176, 182, 205, 234, 239, **4–26**
  see also Murder
Decker, Mrs., 95
Destruction City, Northwest Territories, 47–49, 51, **2–8**
Dewitt, 147
Diamond Tooth Gertie, see Lovejoy, D. T. G.
Dickey, Reverend Robert, 121–22, **5–11**
Divorce, 203, 211, 218, 227, 235, 250
Dogs, 45, 46, 47, 92, 170, 184–85, **3–2, 5–6**
  sledding, 43–44, 45, 48–49, 169–70, 171, 226
Dome City Bank, 213
Donovan Hotel, 221, 223, **8–21**
Dorley, Frances, see Gillis, F. D.
Drange family, 81
Drinking, 115, 125, 172
Drowning, see Death
Dutch Harbor, Alaska, 62
Dyea, Alaska, 69, 108, 113, 169, 189, **4–5** through **4–8, 4–36**
Dyea trail, see Chilkoot trail

E

Earp brothers, 193
Edinburgh, Duke of, **8–25**
Edmonton, Alberta, 29, **2–2**
Edmonton trails, 9, 29–51, 69, 199, 238–39, **I–3, 2–1**
  Overland, 29–34
  Water route, 29, 34–51, 53, **2–3** through **2–7**
Eldorado, **8–1**
Eldorado-Bonanza Quartz and Placer Mining Company, 209
Eldorado creek, 18, 25, 81, 236, 204–06, 207, 211, 217, 220, 233, **8–7, 8–8, 8–10**
Ellingson, Knute, 95
Emilie, 35, 37
Enterprise, 47
Entertainers, 24, 81, 184, 216, 231–32, **I–2, 4–3, 8–26**
Entertainment, 75, 125, 127, 216
Erickson, 92
Excelsior, 3, 19, 180, 205
Excelsior creek, 227

F

Fairbanks, Alaska, 206, 211, 213, 218, 223–24, 234, 235
Fair View hotel, 156–57, 209–11, **6–18, 8–11**
Feero family, 15–17, 24–25, 111–20, 143–45, 147–52, 153–55, 159, 163–66, **1–4, 1–5, 1–6, 5–9, 6–13, 6–27**
  Edith, see Larson, E. F.
  Emma Babcock, 13–19, 113–20, 149–50, 154, 164, **1–6, 6–27**
  Ethel, 13–19, 122–24, **1–4, 1–5, 6–27**

Frank, 16–17, **1–5, 1–6**
John, 14–17, 22–24, 113–20, 147, 149–52, 153–55, 163–66, **1–6**
Stewart, 16–17, 22, **1–5**
Willie, 16–17, 125–26, **1–5**
Fenner, Bert, 95
Field, Mae, 223, 250
Fire, Dawson, 218, 223
Flood, Sheep Camp, 83–85, **4–15, 4–19**
Flora, 59, 174
Flynn, Mr., **4–36**
Food, 25, 35, 36, 41, 47, 69, 81, 102, 149–51, 163–64, 168, 179, 210, 220, **1–9**
Forts
Assiniboine, 34
Cudahy, 217, 245
Grahame, 33
McPherson, 45–47, 48, 51, **2–7**
Resolution, 40, 41, 43, **2–5, 2–6**
Saint John, 33
Selkirk, 169, 188
Wrangell, 190
Yukon, 49–50, 57, **3–4**
Forty Mile Post, Yukon, 205, 217, 245
Fractional claim, 221–22
Fraser, Dan, **4–12**
French family, 146–47, **6–13, 6–14**
Fresno party, see Garner party
Furnace, 211

G

Galvin, Mary Ellen, 169, 201, 217, 245, **8–17**
Pat, 217, 245, **8–17**
Gambling, 81, 127, 129, 155
Garner party, 29–33, **2–2**
G. E., 29, **2–2**
Nellie, 29–33, **2–2**
Gaudet, F. C., 41, 42
Isabella, 41, 48
Gendreau, Frances, see Johndrew, F.
Gilbert, 95
Gillett, Clare M., **4–12**
Gillis, A. J., Dr., 221, **8–20**
Frances Dorley, 176, 186, 198, 220–21, **7–5, 8–20**
Glacial silt, 190
Gold, 22, 152, 154, 163
Gold Bottom creek, 18, 222
Gold Commissioner, 222
Golden belt, 232
Gold Run creek, 219, **8–19**
Gold Run Mining Company, 211
Goldstein, Esther, see Robinson, E. L.
Gould, Hannah S., 61, 235, 240
Graham, Lafayette, 38
Grand Forks, Yukon, 207, 220, **8–10**
Grand Forks Hotel, 207, **8–10**
Griffin, Arthur, 43, 45
Charlie, 43
Grubstaking, 213
Guides, 9, 31, 34–35, 40, 41, 63, 184, 185

H

Hall, Ella, 99–101
Halliday, Doc, 193
Hansen, 105
Hartshorn, Bert, 146
Florence, 108-09, 145–46, 152–53
Hasler, U.S.S., **1–3**
Hawaii, see Sandwich Islands
Hawthorne, Mont, 182
Hayden Brown, 62–63
Healy, Mrs., 69
Healy and Wilson Trading Post, 69, 96
Henderson, Robert, 17
Heney, Michael J., 159–62
Hitchcock, Mary E., 4, 53–61, 209–10, **3–2, 3–6**
Hoffman, Frank, 51
Mrs., 51
Homemaking, 142, 206
Hootalinqua Post, Yukon, 188, 190, 195, 196, **7–14, 7–15**
Hore, T. C., 38
Horn, Cape, 62
Horses, see Pack animals
Hospitals, 193, 223, 235
Hotels, 29, 69, 103, 115, 143–46, 156, 172, 193, 207, 209–11, 217, 218, 219, 221, 234, 235, **5–5, 6–13, 6–18, 8–10, 8–11, 8–19, 8–21**
see also individual names

Hoven, Nettie, 61–63, 235, 240
Hudson's Bay Company, 35, 41, 47
Hunker creek, 95, 235
Hunter, Charles, 194–95, 218–19
Lucille "Ma", 194–95, 218–19
Teslin, 195, 218–19
Hutchison, Ed, **4–12**
Hypothermia, 89

I

Illness, see Sickness
Indians, see Native Americans
Inside Passage, 25–27, 60, 65
Isharov, Marie, see Brady, M. I.
Mr., 95

J

Jacob's Ladder, **4–14**
Johndrew, Frances, 221
Jo Jo Hotel, 219, **8–19**
Joyal Hotel, see Joyelle Hotel
Joyelle Hotel, 235
Judge, Father William, 223
Juneau, Alaska, 22, 120

K

Karn, Mr., 188–89, 196
Kells, Arvin L., 95
Kelly, Ed, family, **4–14**
Kelly, Emma, 96–97, 138, 182–84, 246
Kelly, T. J., 32
King River Falls, 106–08, 179, **4–37**
Klondike discovery, 17–18, 201
Klondike Fever, 204
Klondike Gold Rush park, 65
Klondike Highway, 65, 145
Klondike Kate, see Rockwell, K.
Knudsen family, 81
Kolloen, Erling, 219
Henry, 221, **8–19**
Inga Sjolseth, 72–75, 79–81, 89, 91–92, 96, 105, 172–74, 181, 185–86, 219, **4–11, 8–19**
Koyukuk District, 224–26

L

Ladies of the Maccabees, **6–27**
Ladies Protective Relief Society, 105
Laimee, Marguerite, see Carmack, M. L.
Lakes, 170, 179, 7–7
Bennett, 60, 65, 99, 105, 106, 131, 133, 167, 172, 176, 179, **4–1, 4–2, 4–5, 4–37, 4–38, 4–39, 7–1** through **7–4, 7–6**
Crater, 99–104, **4–2, 4–29, 4–31, 4–32, 4–34**
Deep, 147
Great Slave, 40, 43, 51, **2–5**
LaBerge, 167, 171, 180, 185–88, **4–1, 7–7, 7–13**
LeBarge, see Lake LaBerge
Lindeman, 65, 96, 97, 99, 105, 106, 172, 179, **4–1, 4–2, 4–5, 4–37, 4–38, 7–2, 7–3**
Linderman, see Lake Lindeman
Long, **4–2, 4–35**
Marsh, 180, **4–1**
Tagish, 176, **4–1**
Teslin, 190
Lane, Milley, 186–88
Larrabee, G. W., family, 34
Larson, Edith Feero, 4, 14–17, 23–25, 92–93, 105, 108, 111–20, 143–57, 159–66, **1–1, 1–4, 1–5, 4–35, 5–12, 6–27**
Laundry, 219, 235
Leah, 57–58
Leaming, J. K., 209
Lee, John, **4–12**
Lightering, 69, 113, **5–3, 5–4**
Lilly's feed store, 143
Lippy, claim, 204–05, 233, **8–7**
Salome, 180–82, 201, 204–05
Thomas, 180–82, 204–05
Liquor, see Drinking; or Whiskey
Litigation, 121, 203, 211, 221–22
Little Skookum creek, 221
Log Cabin, British Columbia, White Pass trail, 108, 145, 146, **4–2**
Longuet, V. A., 122
Lovejoy, Diamond Tooth Gertie, **8–26**

Lugard, Flora Shaw, 4, 138–42, 172, 174–76, 245–46, **6–11**
Lynn Canal, 65, 111, 124, 173
Lyons, Esther, see Robinson, E. L.

M

Mackenzie route, see Edmonton trails, Water route
Magic lantern, 54
Maps, **I–3, 2–1, 3–1, 4–1, 4–2, 4–5, 5–8, 6–2, 7–15**
Marriage, 4, 93–95, 189, 203, 206, 211, 219, 224, 227, 234, 235
Marshal, Deputy U.S., 226
Mason, Kate, see Carmack, K. M.
    Skookum Jim, 17–18, 201, **1–7, 1–8**
Matthews, J. V., 108, **4–38**
Mayo District, 218–19
McChord, 198
McConnell, Edward, 234
    Luella Day, Dr., 184–85, 234
McDonald, Alex, 223
McDougall Pass, 48, 49
McKay, J. D. family, 110–11
McNamee, Jim, 223
McPhee, William, 78
McRae, Kate, 240
Meadows, Charley, 82–84, 168–69, 214–16, **4–18**
    Mae McKamish, 4, 22, 82–86, 167–69, 201, 213–16, **4–17, 8–14, 8–15, 8–16**
Melbourne Hotel, 234
Melbourne, Mae, see Meadows, M. M.
Mellen, John F., 63, 235
Melseth, Mr., 81
Menzie, Bob, **4–12**
Miles Canyon rapids, see Rapids, Miles Canyon
Mills, Frank, 235
Mining, 201, 219, 221, **I–4, 8–7, 8–8**
Minna, 189
Missionaries, **4–35**
Monte Cristo hill, 221
Moore, Captain William, 111, 121, 131, **5–1, 5–5**
Moore, Mrs. 177, 220–21
Moosehide, 199, **3–7, 8–4**
Mortimer, Rev. Christopher L., 95
Mosier, Marlin, **4–12**
Mosquitoes, 184, 186, **7–9**
Mothering, 26, 76, 143, 193, 224, 233–34
Mulrooney, Belinda, see Carbonneau, B. M.
    Margaret, 213
Munger, George, 89–91, 98, **4–25**
Murder, 63, 203, 234, 235
Mushing, see Dog sledding
My Ninety Years, 230, 234
My Trip to Alaska in '98, 122

N

Napoleon, 184–85
Native Americans, 11, 17, 32, 36, 40, 41, 47, 48, 50, 55, 65, 68, 69, 75–76, 93, 95, 111, 124, 185, 186, 242, **1–7, 1–8, 2–4, 2–6, 2–7, 4–9, 4–20, 4–33**
Nellie, the pony, 150–52
Nevada Cash Store, 193
Newspapers, 120, 128–29
Nome, Alaska, 49, 214, 234, 235
Northwest Mounted Police, 97, 154, 162, 172, 182
Norton Sound, 54
Nurnberger, William, 95
Nursing, 6, 48, 109–10, 193, 206, 224, 234, 235

O

Ogilvie, William, 17, 76, **1–7**
OK Corral, 193
Optimism, 22, 174
Overland trail, see Chilkoot trail or White Pass trail

P

Pack animals, 22, 29, 30, 32, 69, 73, 85, 86, 96, 113, 119–20, 134–38, 147–52, 168, 176, 189, **4–15, 5–9, 6–7, 6–9, 6–10, 6–12, 6–15**
Packing, 68, 69, 72–76, 86–89, 96–97, 99, 119, 131–41, 147–57, **4–4, 4–9,**

4–10, 4–12 through **4–16, 4–20, 4–21** through **4–24, 4–27, 4–28, 4–30, 4–31, 4–33, 6–1, 6–5** through **6–8**
Palace Grand Theater, 214–16
Paris, France, 203, 211
Partnerships, 72–73, 105, 113, 134, **5–9**
Paulsen, Mr., 81
Pells, **5–9**
Petterson trail, **4–4**
Physicians, 234, **8–20**
Piersoni, Joe, 42
Pilots, see Guides
A Pioneer Woman in Alaska, 34
Polly, see Black, M. M. P.
Porcupine creek, 133, 134, **4–2, 6–15**
Portage, 36, 38, 180, 196, **2–4**
Portland, 19, 22
Prince Napoleon, see Napoleon
Profanity, 129, 138, 149, 171, 226
Professional Men's Boarding House, 221
Prospecting, see Mining
Prospectors' Haven of Retreat, 223
Prostitutes, 69, 141, 145, 150–52, 186–89, 236
Purdy, Lyman, 91, 227, **4–25, 8–24**
  Martha, see Black, M. M. P.
  Will, 89, 227

Q

Queen, 120

R

Railroads, 53, 194
  Alaska, 226, 235
  Canadian Pacific, 29, 230
  Stikine, 194
  Transcontinental, 13, 16, **1–2**
  White Pass and Yukon, 60–61, 69, 132, 143, 157–63, 166, 189, **6–20** through **6–26**
Rainier, Mount, see Tacoma, Mount
Rapids, 36, 180
  Five Finger, 196–198, **8–2**
  Grand, 36, 38
  Miles Canyon, 163, 180–82, 189, **7–8, 7–9**
  One-Mile, see King River Falls
  Rink, 196
  Smith, **2–4**
  Squaw, 182
  White Horse, 163, 180–85, 189, 205, **7–9** through **7–12**
Religion, 22, 73, 147, **4–35, 5–13**
  see also Churches
Restaurants, 69, 82, 102, 143, 146, 190, 207, 220, 221, **4–34**
Rickert, Stacia T. Barnes, 28
Riedeselle, Marie, 33
Rio de Janeiro, Brazil, 61
Rivers
  Athabasca, 29
  Dyea, 69, 78, **I–2, 4–6, 4–9, 4–14**
  Forty Mile, 217
  Fox, 33
  Hay, 43
  Klondike, 8, 234, **8–23**
  Mackenzie, 29, 44, 45, 53
  Peace, 33, **7–15**
  Peel, 29, 45
  Pelly, 188, 198
  Porcupine, 47
  Rat, 29, 46, 47, 49
  Root, 45
  Skagway, 111, 133, **6–3, 6–10**
  Slave, 29, 42, **2–3, 2–4**
  Stewart, 198, **8–3**
  Stikine, 190, 194, **7–15**
  Taiya, see Dyea
  Teslin, 190, 194, **7–14**
  Thirty Mile, 186–190, **7–13**
  White, 198
  Yukon, 29, 47, 53, 58–59, 65, 69, 190, 196–199, **3–3** through **3–6, 7–14, 8–1, 8–2, 8–3**
Rockwell, Kate, 184
Robinson, Esther Lyons, 67–68, 231–33, 242, **4–3, 8–26**
Romig, Emily Craig, 4, 34–51, 235, **2–8**
  J. H., Dr., 235
Rosie, 51
Royal Colonial Institute, 140, 141
Russel, 221, 222
Ryne, Kate, 109–10, 243

S

Sadlemire, Mr., 189
Saint Albert, Alberta, 29
Saint Elias mountains, 167
Saint Mary's Hospital, 223
Saint Michaels, Alaska, 54–55, 56, 62, 63, 210, **3–3**
Saint Paul, 54, 56, **3–2**
Saloons, see Bars
Sandwich Islands, 89, 227
Sanvig, 73, 105, 172–74, 185–86
Savoy Theater, see Palace Grand Theater
Sawmill, 167, 182, 227, **4–2, 5–5**
Scales, Chilkoot trail, 93, **4–2**
Scandinavian-American Bank, 219
Schools, 121–24, **5–12**
Scows, 176, 182, 205, **2–5, 7–10, 7–11, 8–2**
Scurvy, 33, 48, 193
Seattle, Washington, 62, 201–02, **1–11**
Seattle No. 1, **3–7**
Seigfried, 95
Senkler, E. C., Gold Commissioner, 222
Service, Robert, 185
Seward, Alaska, 226, 235
Shaw, Flora, see Lugard, F. S.
Sheep Camp, Alaska, 78–85, 96, **4–2, 4–5, 4–16, 4–19**
Shell game, **6–17**
"She's Such a Nice Girl, Too", 232
Shooting, 216
Sickness, 48, 57, 73–75, 106, 108, 109–10, 164–65, 201, 230, 234, 235
Sjolseth, Inga, see Kolloen, I. S.
Skagua, 124
Skaguay, see Skagway
Skaguay *News*, 120
Skagway Belle, 150–52, 244
Skagway, Alaska, 13, 24, 60, 111–29, 124, 134, 189, **1–12, 5–1** through **5–15, 6–2, 6–21, 6–22**
Skagway trail, see White Pass trail
Skookum, 18
Skookum Jim, see Mason, S. J.
Smith, Jefferson, 154–57, 209
Smith, Soapy, see Smith, Jefferson
Snowslide, see Avalanche
Soap creek, 222
Sovereign, 63
Spencer, Captain, 103, 184
Split-ups, 41, 43, 105
Sporting women, see Prostitutes
Springer, R. M., 37, 38
Stampede, 8–12, 21–24, 32, **1–9, 1–10, 1–11**
Stealing, see Crime
Steamers, 24–25, 57–59, 172, 174, 189–90, 196, **2–5, 3–4, 3–5, 3–7, 7–14, 8–1, 8–2**
  see also individual names
Steele, Sam, 182, 184
Steffens, Gus, 95
Stevenson, W. A., 95
Stewart, A. D., 51, 240
Stewart, Arora Susie, 189–90
  C. J., 189
Stewart Company, 189
Stikeen, see Stikine
Stikine trail, 9, 109, 190–95, **I–3, 7–15**
Stone House, Chilkoot trail, 95, 96, **4–2**
Straits of Magellan, 62
Stresses, 104–10, 170–71
Stringer, Mrs., 47
Stroller, see White, E. J.
Strong, Annie Hall, 120–21
Suicide, 109, 149, **4–38**
Suing, see Litigation
Surveying, 221, 222, **1–7**

T

Tacoma Hotel, 102
Tacoma, Mount, 13, **1–3**
Tacoma, Washington, 15–16, **1–2, 1–3, 1–10**
Tagish Charley, **1–8**
Tailing, 154
Tanana District, see Fairbanks, Alaska
Teachers, 122, **5–12**
Telegraph Creek, British Columbia, 109–10, 190
Tenting, 45, 82, 83–85, 91, 115–19, **2–6, 2–7, 4–13, 4–34, 4–35, 4–39, 5–2, 5–6, 5–7**
Teslin, British Columbia, 195
Thoerner, Fred, 235
Thompson, Alfred, Dr., **8–20**

Tiburon Island, Mexico, 216
Tlingits, 65, 111
"Today's the Day They Give Babies Away", 24
Tombstone, Arizona, 193
Tourists, 53
Tracking, 45, 47
The Tragedy of the Klondike, 184, 234
Tram, 86, 95, 132, 143, **4–28, 7–9**
Tunnel Mountain, White Pass and Yukon Railroad, **6–23**
Turkeys, 152–53, **6–16**
Tutchone, Southern, 185
Two Women in the Klondike, 54

U

Unalaska, Alaska, 54
Union Church, 121–22, 125, **5–11, 5–12, 5–13**

V

Valdez, Alaska, 28
Valparaiso, Chili, 62
Van Buren, Edith, 4, 53–61, **3–2**
Victoria, British Columbia, 28, 227
Voorhees, (Bert?), 188–89, 196

W

Waldal, Mr., 81
Ward, Phil, 93
Warmolts, Lambertus, 34–35, 41
Water Company, 211
Wedding, see Marriage
Wharves, 69, 113, 121, **5–4**
Whipsawing, 44, 49, 167, **7–1**
Whiskey, 153–54, 156
Whitechapel, 188
White, E. J. "Stroller", 128–29, 194
White, Georgia Hacker, 4, 105–06, 188–89, 196, 235, **4–36**
White Horse rapids, see Rapids, White Horse
Whitehorse, Yukon, 163, 185, 189, 230, **8–25**
White Pass, 60, 111, 132, 133, 145–46, 162–63, **4–2, 6–2, 6–5, 6–6, 6–7, 6–24, 6–25**
White Pass City, Alaska, 60, 133, 143–45, 146–47, 159, 162, **6–5, 6–13**
White Pass hotel, 143, **6–13**
White Pass trail, 9, 65, 69, 108, 111, 131–57, 163–66, **I–3, 4–2, 6–1** through **6–10, 6–12** through **6–17, 6–19, 6–25**
Widows, 6, 27, 61, 95, 128, 213, 235
Wild Horses and Gold, 239
Willamette, **1–10, 1–11**
Williams, **4–12**
Willis, Mrs. J. T., 4
Wilson, Cad, see Robinson, E. L.
Wilson, Josephine, 68
  Veazie, 67–68, 231
Wilson, Mrs., 184–85
Writing, 26–27, 56, 236

Y

Yakima, Washington, 213, **8–13**
Yukon government, 230, 234
Yukon infantry, 230
Yukon River route, 4, 9, 52–63, 69, **I–3, 3–1**
Yukon Territory, 9
Yuma, Arizona, 194, 216, **8–15, 8–16**

A Note about the Author

Melanie J. Mayer, Ph.D., is Professor of Psychology and Psychobiology at the Santa Cruz campus of the University of California. She is author of *Demonstrations of Sensory Perception* (Wiley, 1982), as well as of numerous research papers. Her enjoyment of hiking, backpacking, camping, and cross-country skiing complements her interests in the wilderness, in conservation, and in North American prehistory and history.